STATA CLUSTER ANALYSIS
REFERENCE MANUAL
RELEASE 8

A Stata Press Publication
STATA CORPORATION
College Station, Texas

Stata Press, 4905 Lakeway Drive, College Station, Texas 77845

The suggested citation for this software is

StataCorp. 2003. *Stata Statistical Software: Release 8.0*. College Station, TX: Stata Corporation.

Table of Contents

Cross-Referencing the Documentation

When reading this manual, you will find references to other Stata manuals. For example,

[U] **29 Overview of Stata estimation commands**

[R] **regress**

[P] **matrix define**

The first is a reference to Chapter 29, *Overview of Stata estimation commands* in the *Stata User's Guide*, the second is a reference to the `regress` entry in the *Base Reference Manual*, and the third is a reference to the `matrix define` entry in the *Programming Reference Manual*.

All of the manuals in the Stata Documentation have a shorthand notation, such as [U] for the *User's Guide* and [R] for the *Base Reference Manual*.

The complete list of shorthand notations and manuals is as follows:

[GSM]	*Getting Started with Stata for Macintosh*
[GSU]	*Getting Started with Stata for Unix*
[GSW]	*Getting Started with Stata for Windows*
[U]	*Stata User's Guide*
[R]	*Stata Base Reference Manual*
[G]	*Stata Graphics Reference Manual*
[P]	*Stata Programming Reference Manual*
[CL]	*Stata Cluster Analysis Reference Manual*
[XT]	*Stata Cross-Sectional Time-Series Reference Manual*
[SVY]	*Stata Survey Data Reference Manual*
[ST]	*Stata Survival Analysis & Epidemiological Tables Reference Manual*
[TS]	*Stata Time-Series Reference Manual*

Detailed information about each of these manuals may be found online at

http://www.stata-press.com/manuals/

Title

> **intro** — Introduction to cluster analysis manual

Description

This entry describes the *Stata Cluster Analysis Reference Manual*.

Remarks

This manual documents the `cluster` command, and is referred to as [CL] in cross-references. Following this entry, [CL] **cluster** provides an overview of the `cluster` command.

This manual is arranged alphabetically. If you are new to Stata's cluster analysis commands, we recommend that you read the following sections first:

[CL] **cluster**	Introduction to cluster analysis commands
[CL] **cluster kmeans**	Kmeans cluster analysis
[CL] **cluster singlelinkage**	Single linkage cluster analysis
[CL] **cluster dendrogram**	Dendrograms for hierarchical cluster analysis
[CL] **cluster stop**	Cluster analysis stopping rules
[CL] **cluster generate**	Generate summary or grouping variables from a cluster analysis

Stata is continually being updated. Stata users are always writing new commands, as well. To find out about the latest cluster analysis features, type `search cluster analysis` after installing the latest official updates; see [R] **update**.

Four different datasets are used in the examples in this manual. They can be accessed by typing one of the following:

```
use http://www.stata-press.com/data/r8/labtech
use http://www.stata-press.com/data/r8/homework
use http://www.stata-press.com/data/r8/physed
use http://www.stata-press.com/data/r8/wclub
```

What's new

Stata 8 has four new hierarchical cluster analysis methods, the addition of two cluster stopping rules, two new distance measures, and added programming features. This section is intended for previous Stata users. If you are new to Stata, you may as well skip it.

1. The new `cluster wardslinkage` command provides Ward's linkage hierarchical cluster analysis. The well-known Ward's method (also known as minimum-variance clustering) is obtained by default. Ward's method combines groups in the hierarchy based on the minimization of squared error. See [CL] **cluster wardslinkage**.

2. The new `cluster waveragelinkage` command provides weighted-average linkage hierarchical cluster analysis. Weighted-average linkage is a variation on average linkage clustering (which is available with the `cluster averagelinkage` command; see [CL] **cluster averagelinkage**). Weighted-average linkage gives equal weight to groups being combined in the hierarchy, regardless of the number of observations in each group. Average linkage, on the other hand, gives each observation equal weight when combining groups, meaning that groups with more observations will have more influence on the determination of the combined group. See [CL] **cluster waveragelinkage**.

1

3. The new `cluster centroidlinkage` command provides centroid linkage hierarchical cluster analysis. Centroid linkage groups observations into a hierarchy. The groups are combined together based on the distance between the centroids (means) of the groups. This differs from the `cluster averagelinkage` command, which combines groups based on the average of the distances between observations of the two groups to be combined. See [CL] **cluster centroidlinkage**.

4. The new `cluster medianlinkage` command provides median linkage hierarchical cluster analysis. Median linkage (also known as Gower's method) is a variation on centroid linkage. Median linkage gives equal weight to groups being combined in the hierarchy, regardless of the number of observations in each group. Centroid linkage, on the other hand, gives each observation equal weight when combining groups, meaning that groups with more observations will have more influence on the determination of the combined group. See [CL] **cluster medianlinkage**.

5. The new `cluster stop` command provides cluster stopping rules. Stopping rules help answer the following question: How many clusters are there? Two popular stopping rules are provided, the Caliński & Harabasz pseudo-F index and the Duda & Hart Je(2)/Je(1) index with associated pseudo T-squared. See [CL] **cluster stop**.

6. Additional stopping rules are easily added to the new `cluster stop` command. See [CL] **cluster programming subroutines**.

7. Two new dissimilarity measures have been added, L2squared and Lpower(#). L2squared provides squared Euclidean distance. Lpower(#) provides the Minkowski distance metric with argument # raised to the # power. See *Similarity and dissimilarity measures* in [CL] **cluster**.

8. A list of the variables used in the cluster analysis is now saved with the cluster analysis structure. This can be seen using the `cluster list` command (see [CL] **cluster utility**), and can be accessed by programmers using the `cluster query` command (see [CL] **cluster programming utilities**). This will be helpful to programmers adding new features to the `cluster` command; see [CL] **cluster programming subroutines**.

For a complete list of all the new features in Stata 8, see [U] **1.3 What's new**.

Also See

Background: [R] **intro**

Title

cluster — Introduction to cluster analysis commands

Syntax

cluster *subcommand* ...

Description

Stata's cluster analysis routines give you a choice of several hierarchical and partition clustering methods. Post-clustering summarization methods and cluster management tools are also provided. This entry presents an overview of cluster analysis, the `cluster` command, and Stata's cluster analysis management tools. The similarity and dissimilarity measures available for use with the cluster analysis methods are also explained.

The `cluster` command has the following *subcommand*s, which are detailed in their respective manual entries.

Partition clustering methods

kmeans	[CL] **cluster kmeans**	Kmeans cluster analysis
kmedians	[CL] **cluster kmedians**	Kmedians cluster analysis

Hierarchical clustering methods

singlelinkage	[CL] **cluster singlelinkage**	Single linkage cluster analysis
averagelinkage	[CL] **cluster averagelinkage**	Average linkage cluster analysis
completelinkage	[CL] **cluster completelinkage**	Complete linkage cluster analysis
waveragelinkage	[CL] **cluster waveragelinkage**	Weighted-average linkage cluster analysis
medianlinkage	[CL] **cluster medianlinkage**	Median linkage cluster analysis
centroidlinkage	[CL] **cluster centroidlinkage**	Centroid linkage cluster analysis
wardslinkage	[CL] **cluster wardslinkage**	Ward's linkage cluster analysis

Post-clustering commands

stop	[CL] **cluster stop**	Cluster analysis stopping rules
dendrogram	[CL] **cluster dendrogram**	Dendrograms for hierarchical cluster analysis
generate	[CL] **cluster generate**	Generate summary or grouping variables from a cluster analysis

User utilities

notes	[CL] **cluster notes**	Place notes in cluster analysis
dir	[CL] **cluster utility**	Directory list of cluster analyses
list	[CL] **cluster utility**	List cluster analyses
drop	[CL] **cluster utility**	Drop cluster analyses
rename	[CL] **cluster utility**	Rename cluster analyses
renamevar	[CL] **cluster utility**	Rename cluster analysis variables

3

Programmer utilities

	[CL] **cluster programming subroutines**	Add cluster analysis routines
query	[CL] **cluster programming utilities**	Obtain cluster analysis attributes
set	[CL] **cluster programming utilities**	Set cluster analysis attributes
delete	[CL] **cluster programming utilities**	Delete cluster analysis attributes
parsedistance	[CL] **cluster programming utilities**	Parse (dis)similarity measure names
measures	[CL] **cluster programming utilities**	Compute (dis)similarity measures

Remarks

Remarks are presented under the headings

> Introduction to cluster analysis
> Stata's cluster analysis system
> Data transformations and variable selection
> Similarity and dissimilarity measures
>> Similarity and dissimilarity measures for continuous data
>> Similarity measures for binary data
>> Binary similarity measures applied to averages
> Partition cluster analysis methods
> Hierarchical cluster analysis methods
>> Agglomerative methods
>> Lance and Williams' recurrence formula
>> (Dis)similarity transformations and the Lance and Williams formula
>> Warning concerning (dis)similarity choice
>> Synonyms
>> Reversals
> Post-clustering commands
> Cluster management tools

Introduction to cluster analysis

Cluster analysis attempts to determine the natural groupings (or clusters) of observations. Sometimes it is called "classification", but this term is used by others to mean discriminant analysis, which, although related to cluster analysis, is not the same. To avoid confusion, we will use "cluster analysis" or "clustering" when referring to finding groups in data. It is difficult (maybe impossible) to give a definition of cluster analysis. Kaufman and Rousseeuw (1990) start their book by saying, "Cluster analysis is the art of finding groups in data." Everitt, Landau, and Leese (2001, 6) use the terms "cluster", "group", and "class", and say concerning a formal definition for these terms that, "In fact it turns out that such formal definition is not only difficult but may even be misplaced."

Who uses cluster analysis and why? Everitt, Landau, and Leese (2001) and Gordon (1999) provide examples of the use of cluster analysis. These include the refining or redefining of diagnostic categories in psychiatry, the detection of similarities in artifacts by archaeologists to study the spatial distribution of artifact types, the discovery of hierarchical relationships in the field of taxonomy, and the identification of sets of similar cities so that one city from each class can be sampled in a market research task. In addition, the activity that is now called "data mining" relies extensively on cluster analysis methods.

We view cluster analysis as an exploratory data analysis technique. This view is shared by Everitt (1993, 10). He says, speaking of cluster analysis techniques, "Many of these have taken their place alongside other exploratory data analysis techniques as tools of the applied statistician. The term exploratory is important here since it explains the largely absent 'p-value', ubiquitous in many other areas of statistics." He then says, "clustering methods are intended largely for generating rather than testing hypotheses." This states the case very well.

It has been said that there are as many cluster analysis methods as there are people performing cluster analysis. This is a gross understatement! There exist infinitely more ways to perform a cluster analysis than people who perform them.

There are several general types of cluster analysis methods, and within each of these there are numerous specific methods. Additionally, most cluster analysis methods allow a variety of distance measures for determining the similarity or dissimilarity between observations. Some of the measures do not meet the requirements to be called a distance metric, so the more general term "dissimilarity measure" is used in place of distance. Similarity measures may also be used in place of dissimilarity measures. There are an infinite number of (dis)similarity measures. For instance, there are an infinite number of Minkowski distance metrics, with the familiar Euclidean, absolute-value, and maximum-value distances being special cases.

In addition to cluster method and (dis)similarity measure choice, someone performing a cluster analysis might decide to perform data transformations and/or variable selection before clustering. Then, there is the determining of how many clusters there really are in the data. Stopping rules are used to help determine the number of clusters. There is a surprisingly large number of stopping rules mentioned in the literature. For example, Milligan and Cooper (1985) compare 30 different stopping rules.

Looking at all of these choices, you can see why there are more cluster analysis methods than people performing cluster analysis.

Stata's cluster analysis system

Stata's `cluster` command was designed to allow you to keep track of the various cluster analyses performed on your data. The main clustering subcommands `singlelinkage`, `averagelinkage`, `completelinkage`, `waveragelinkage`, `medianlinkage`, `centroidlinkage`, `wardslinkage`, `kmeans`, and `kmedians` create named Stata cluster objects that keep track of the variables these methods create and hold other identifying information for the cluster analysis. These cluster objects become part of your dataset. They are saved with your data when your data are `saved`, and are retrieved when you again `use` your dataset; see [R] **save**.

Post-cluster analysis subcommands are available with the `cluster` command to help examine the created clusters. Cluster management tools are provided that allow you to add information to the cluster objects and to manipulate them as needed.

The main clustering subcommands, available similarity and dissimilarity measures, post-clustering subcommands, and cluster management tools are discussed in the following sections. Stata's cluster analysis system is extendable in many ways. Programmers wishing to add to the cluster system should see [CL] **cluster programming subroutines**.

Stata's clustering methods fall into two general types: partition and hierarchical. These two types are discussed below. There exist other types, such as fuzzy partition (where observations can belong to more than one group). Stata's `cluster` command is designed so that programmers can add more methods of whatever type they desire; see [CL] **cluster programming subroutines** and [CL] **cluster programming utilities** for details.

❏ Technical Note

For those familiar with Stata's large array of estimation commands, we warn you not to get confused between cluster analysis (the `cluster` command) and the `cluster()` option allowed with many estimation commands. Cluster analysis finds groups in data. The `cluster()` option allowed with various estimation commands indicates that the observations are independent across the groups defined by the option, but not necessarily independent within those groups. A grouping variable produced by the `cluster` command will seldom satisfy the assumption behind the use of the `cluster()` option.

❏

Data transformations and variable selection

Stata's `cluster` command does not have any built-in data transformations, but, since Stata has full data management and statistical capabilities, you can use other Stata commands to transform your data before calling the `cluster` command. In some cases, standardization of the variables is important to keep a variable with high variability from dominating the cluster analysis. In other cases, standardization of variables acts to hide the true groupings present in the data. The decision to standardize or perform other data transformations depends heavily on the type of data that you are analyzing and on the nature of the groups that you are trying to discover.

A related topic is the selection of variables to use in the cluster analysis. Data transformations (such as standardization of variables) and the variables selected for use in clustering can have a large impact on the groupings that are discovered. These and other cluster analysis data issues are covered in many of the cluster analysis texts, including Anderberg (1973), Gordon (1999), Everitt, Landau, and Leese (2001), and Späth (1980). Milligan and Cooper (1988) and Schaffer and Green (1996) are sources completely devoted to the issue of data transformation associated with cluster analysis.

Similarity and dissimilarity measures

A variety of similarity and dissimilarity measures have been implemented for Stata's clustering commands. Some of these measures were designed for continuous variables, while others were designed for binary variables. In the formulas below, x_{ab} is the value of variable a and observation b. All summations and maximums are over the p variables involved in the cluster analysis for the two observations in the (dis)similarity comparison. Do not confuse this with most other contexts, where the summations and maximums are over the observations. For clustering, we compare two observations across their variables.

Similarity and dissimilarity measures for continuous data

The similarity and dissimilarity measures for continuous data available in Stata include the following:

L2 (alias <u>Euc</u>lidean)
requests the Minkowski distance metric with argument 2

$$\left\{ \sum_{k=1}^{p} (x_{ki} - x_{kj})^2 \right\}^{1/2}$$

which is best known as Euclidean distance. This is the default dissimilarity measure for all the `cluster` subcommands except for `centroidlinkage`, `medianlinkage`, and `wardslinkage`, which default to using L2squared.

L2squared

requests the square of the Minkowski distance metric with argument 2

$$\sum_{k=1}^{p}(x_{ki} - x_{kj})^2$$

which is best known as squared Euclidean distance. This is the default dissimilarity measure for the `centroidlinkage`, `medianlinkage`, and `wardslinkage` subcommands of `cluster`.

L1 (aliases <u>abs</u>olute, `cityblock`, and <u>manhat</u>tan)

requests the Minkowski distance metric with argument 1

$$\sum_{k=1}^{p}|x_{ki} - x_{kj}|$$

which is best known as absolute-value distance.

<u>Linf</u>inity (alias <u>maxim</u>um)

requests the Minkowski distance metric with infinite argument

$$\max_{k=1,\dots,p}|x_{ki} - x_{kj}|$$

and is best known as maximum-value distance.

L(#)

requests the Minkowski distance metric with argument #:

$$\left(\sum_{k=1}^{p}|x_{ki} - x_{kj}|^{\#}\right)^{1/\#} \qquad \# \geq 1$$

We discourage the use of extremely large values for #. Since the absolute value of the difference is being raised to the value of #, depending on the nature of your data, you could experience numeric overflow or underflow. With a large value of #, the L() option will produce similar cluster results to the `Linfinity` option. Use the numerically more stable `Linfinity` option instead of a large value for # in the L() option.

See Anderberg (1973) for a discussion of the Minkowski metric and its special cases.

Lpower(#)

requests the Minkowski distance metric with argument #, raised to the # power:

$$\sum_{k=1}^{p}|x_{ki} - x_{kj}|^{\#} \qquad \# \geq 1$$

As with L(#), we discourage the use of extremely large values for #; see the discussion above.

Canberra
  requests the following distance metric

$$\sum_{k=1}^{p} \frac{|x_{ki} - x_{kj}|}{|x_{ki}| + |x_{kj}|}$$

  which takes values between 0 and p, the number of variables used in the cluster analysis; see Gordon (1999) and Gower (1985). Gordon (1999) explains that the Canberra distance is very sensitive to small changes near zero.

correlation
  requests the correlation coefficient similarity measure

$$\frac{\sum_{k=1}^{p}(x_{ki} - \overline{x}_{.i})(x_{kj} - \overline{x}_{.j})}{\left\{ \sum_{k=1}^{p}(x_{ki} - \overline{x}_{.i})^2 \sum_{l=1}^{p}(x_{lj} - \overline{x}_{.j})^2 \right\}^{1/2}}$$

  where $\overline{x}_{.j}$ is the mean for observation j over the p variables in the cluster analysis.

  The correlation similarity measure takes values between -1 and 1. With this measure, the relative direction of the two observation vectors is important. The correlation similarity measure is related to the angular separation similarity measure (described next). The correlation similarity measure gives the cosine of the angle between the two observation vectors measured from the mean; see Gordon (1999).

angular (alias angle)
  requests the angular separation similarity measure

$$\frac{\sum_{k=1}^{p} x_{ki} x_{kj}}{\left(\sum_{k=1}^{p} x_{ki}^2 \sum_{l=1}^{p} x_{lj}^2 \right)^{1/2}}$$

  which is the cosine of the angle between the two observation vectors measured from zero, and takes values from -1 to 1; see Gordon (1999).

Similarity measures for binary data

  Similarity measures for binary data are based on the four values from the cross-tabulation of the two observations.

		obs. j	
		1	0
obs.	1	a	b
i	0	c	d

  a is the number of variables where observations i and j both had ones, and d is the number of variables where observations i and j both had zeros. The number of variables where observation i is one and observation j is zero is b, and the number of variables where observation i is zero and observation j is one is c.

  The cluster command follows Stata's general practice of treating nonzero values as a one when a binary variable is expected. Specifying one of the binary similarity measures imposes this behavior.

Gower (1985) gives an extensive list of fifteen binary similarity measures. Fourteen of these are implemented in Stata. (The excluded measure has many cases where the quantity is undefined and so was not implemented.) Anderberg (1973) gives an interesting table where many of these measures are compared based on whether the zero–zero matches are included in the numerator, whether these matches are included in the denominator, and how the weighting of matches and mismatches is handled. Hilbe (1992a, 1992b) implemented an early Stata command for computing some of these (as well as other) binary similarity measures.

The formulas for some of these binary similarity measures are undefined when either one or both of the observations are all zeros (or, in some cases, all ones). Gower (1985) says concerning these cases, "These coefficients are then conventionally assigned some appropriate value, usually zero."

The following binary similarity coefficients are available in the `cluster` command. Unless stated otherwise, the similarity measures range from 0 to 1.

matching
: requests the simple matching (Sokal and Michener 1958) binary similarity coefficient

$$\frac{a + d}{a + b + c + d}$$

which is the proportion of matches between the two observations.

Jaccard
: requests the Jaccard (1908) binary similarity coefficient

$$\frac{a}{a + b + c}$$

which is the proportion of matches when at least one of the observations had a one. If both observations are all zeros, this measure is undefined. In this case, Stata declares the answer to be one, meaning perfect agreement. This is a reasonable choice for cluster analysis, and will cause an all-zero observation to have similarity of one only with another all-zero observation. In all other cases, an all-zero observation will have Jaccard similarity of zero to the other observation.

Russell
: requests the Russell & Rao (1940) binary similarity coefficient

$$\frac{a}{a + b + c + d}$$

Hamman
: requests the Hamman (1961) binary similarity coefficient

$$\frac{(a + d) - (b + c)}{a + b + c + d}$$

which is the number of agreements minus disagreements divided by the total. The Hamman coefficient ranges from -1, perfect disagreement, to 1, perfect agreement. The Hamman coefficient is equal to twice the simple matching coefficient minus 1.

Dice
>   requests the Dice binary similarity coefficient

$$\frac{2a}{2a+b+c}$$

suggested by Czekanowski (1932), Dice (1945), and Sørensen (1948). The Dice coefficient is similar to the Jaccard similarity coefficient, but gives twice the weight to agreements. Like the Jaccard coefficient, the Dice coefficient is declared by Stata to be one if both observations are all zero, thus avoiding the case where the formula is undefined.

antiDice
>   requests the binary similarity coefficient

$$\frac{a}{a+2(b+c)}$$

which is credited to Anderberg (1973). We did not call this the Anderberg coefficient, since there is another coefficient better known by that name; see the `Anderberg` option. The name antiDice is our creation. This coefficient takes the opposite view from the Dice coefficient and gives double weight to disagreements. As with the Jaccard and Dice coefficients, the antiDice coefficient is declared to be one if both observations are all zeros.

Sneath
>   requests the Sneath & Sokal (1962) binary similarity coefficient

$$\frac{2(a+d)}{2(a+d)+(b+c)}$$

which is similar to the simple matching coefficient, but gives double weight to matches. Also compare the Sneath & Sokal coefficient with the Dice coefficient, which differs only in whether it includes d.

Rogers
>   requests the Rogers & Tanimoto (1960) binary similarity coefficient

$$\frac{a+d}{(a+d)+2(b+c)}$$

which takes the opposite approach from the Sneath & Sokal coefficient and gives double weight to disagreements. Also compare the Rogers & Tanimoto coefficient with the antiDice coefficient, which differs only in whether it includes d.

Ochiai
>   requests the Ochiai (1957) binary similarity coefficient

$$\frac{a}{\left\{(a+b)(a+c)\right\}^{1/2}}$$

The formula for the Ochiai coefficient is undefined when one, or both, of the observations being compared is all zeros. If both are all zeros, Stata declares the measure to be one, and if only one of the two observations is all zeros, the measure is declared to be zero.

Yule

requests the Yule (see Yule and Kendall 1950) binary similarity coefficient

$$\frac{ad - bc}{ad + bc}$$

which ranges from -1 to 1. The formula for the Yule coefficient is undefined when one or both of the observations are either all zeros or all ones. Stata declares the measure to be 1 when $b+c=0$, meaning there is complete agreement. Stata declares the measure to be -1 when $a+d=0$, meaning there is complete disagreement. Otherwise, if $ad-bc=0$, Stata declares the measure to be 0. These rules, applied before using the Yule formula, avoid the cases where the formula would produce an undefined result.

Anderberg

requests the Anderberg binary similarity coefficient

$$\left(\frac{a}{a+b} + \frac{a}{a+c} + \frac{d}{c+d} + \frac{d}{b+d}\right)\Big/4$$

The Anderberg coefficient is undefined when one or both observations are either all zeros or all ones. This difficulty is overcome by first applying the rule that if both observations are all ones (or both observations are all zeros), then the similarity measure is declared to be one. Otherwise, if any of the marginal totals ($a+b$, $a+c$, $c+d$, $b+d$) are zero, then the similarity measure is declared to be zero.

Kulczynski

requests the Kulczynski (1927) binary similarity coefficient

$$\left(\frac{a}{a+b} + \frac{a}{a+c}\right)\Big/2$$

The formula for this measure is undefined when one or both of the observations are all zeros. If both observations are all zeros, Stata declares the similarity measure to be one. If only one of the observations is all zeros, the similarity measure is declared to be zero.

Gower2

requests the binary similarity coefficient

$$\frac{ad}{\{(a+b)(a+c)(d+b)(d+c)\}^{1/2}}$$

which, presumably, was first presented by Gower (1985). Stata uses the name Gower2 to avoid confusion with the better-known Gower coefficient (not currently in Stata), which is used to combine continuous and categorical (dis)similarity measures computed on a dataset into one measure.

The formula for this similarity measure is undefined when one or both of the observations are all zeros or all ones. This is overcome by first applying the rule that if both observations are all ones (or both observations are all zeros), then the similarity measure is declared to be one. Otherwise, if $ad=0$, then the similarity measure is declared to be zero.

Pearson
   requests Pearson's ϕ (see Guilford 1942) binary similarity coefficient

$$\frac{ad - bc}{\left\{(a+b)(a+c)(d+b)(d+c)\right\}^{1/2}}$$

   which ranges from -1 to 1. The formula for this coefficient is undefined when one or both of the observations are either all zeros or all ones. Stata declares the measure to be 1 when $b + c = 0$, meaning there is complete agreement. Stata declares the measure to be -1 when $a + d = 0$, meaning there is complete disagreement. Otherwise, if $ad - bc = 0$, Stata declares the measure to be 0. These rules, applied before using Pearson's ϕ coefficient formula, avoid the cases where the formula would produce an undefined result.

Binary similarity measures applied to averages

   Some cluster analysis methods (such as Stata's kmeans and kmedians clustering) need to compute the (dis)similarity between observations and group averages or group medians. With binary data, a group average is interpreted as a proportion.

   A group median for binary data will be zero or one, except when there are an equal number of zeros and ones. In this case, Stata calls the median 0.5, which can also be interpreted as a proportion.

   In Stata's `cluster kmeans` and `cluster kmedians` commands for the case of comparing a binary observation to a group proportion, (see *Partition cluster analysis methods*), the values of a, b, c, and d are obtained by assigning the appropriate fraction of the count to these values. In our earlier table showing the relationship of a, b, c, and d in the cross-tabulation of observation i and observation j, if we replace observation j by the group-proportions vector, then when observation i is 1 we add the corresponding proportion to a, and add one minus that proportion to b. When observation i is 0, we add the corresponding proportion to c, and add one minus that proportion to d. After the values of a, b, c, and d are computed in this way, the binary similarity measures are computed using the formulas as already described.

Partition cluster analysis methods

   Partition methods break the observations into a distinct number of nonoverlapping groups. There are many different partition methods. Stata has implemented two of them, kmeans and kmedians.

   One of the more commonly used partition clustering methods is called kmeans cluster analysis. In kmeans clustering, the user specifies the number of clusters to create. These k clusters are formed by an iterative process. Each observation is assigned to the group whose mean is closest, and then based on that categorization, new group means are determined. These steps continue until no observations change groups. The algorithm begins with k seed values, which act as the k group means. There are many ways to specify the beginning seed values. See [CL] **cluster kmeans** for the details of the `cluster kmeans` command.

   A variation of kmeans clustering is kmedians clustering. The same process is followed in kmedians as in kmeans, with the exception that medians instead of means are computed to represent the group centers at each step; see [CL] **cluster kmedians** for details.

   These partition clustering methods will generally be quicker, and will allow larger datasets than the hierarchical clustering methods outlined next. However, if you wish to examine clustering to various numbers of clusters, you will need to execute `cluster` numerous times with the partition methods. Clustering to various numbers of groups using a partition method will typically not produce clusters that are hierarchically related. If this is important for your application, consider using one of the hierarchical methods.

Hierarchical cluster analysis methods

Hierarchical clustering methods are generally of two types: agglomerative or divisive. Hierarchical clustering creates (by either dividing or combining) hierarchically related sets of clusters.

Agglomerative hierarchical clustering methods begin with each observation being considered as a separate group (N groups each of size 1). The closest two groups are combined ($N - 1$ groups, one of size 2 and the rest of size 1), and this process continues until all observations belong to the same group. This process creates a hierarchy of clusters.

In addition to choosing the similarity or dissimilarity measure to use in comparing two observations, there is the choice of what should be compared between groups that contain more than one observation. The method of comparing groups is called a linkage method. Stata's `cluster` command provides several hierarchical agglomerative linkage methods, which are discussed in the next section.

Unlike the hierarchical agglomerative clustering methods, in divisive hierarchical clustering, you begin with all observations belonging to one group. This group is then split in some fashion to create two groups. One of these two groups is then split to create three groups. One of these three groups is split to create four groups, and so on, until all observations are in their own separate group. Stata does not currently have any divisive hierarchical clustering commands. There are relatively few mentioned in the literature, and they tend to be particularly time-consuming to compute.

To appreciate the underlying computational complexity of both agglomerative and divisive hierarchical clustering, consider the following information paraphrased from Kaufman and Rousseeuw (1990). The first step of an agglomerative algorithm considers $N(N-1)/2$ possible fusions of observations to find the closest pair. This number grows quadratically with N. For divisive hierarchical clustering, the first step would attempt to find the best split into two nonempty subsets, and if all possibilities were considered, it would amount to $2^{(N-1)} - 1$ comparisons. This number grows exponentially with N.

Agglomerative methods

The hierarchical agglomerative linkage methods provided by Stata's `cluster` command are single linkage, complete linkage, average linkage, Ward's method, centroid linkage, median linkage, and weighted-average linkage. There are others mentioned in the literature, but these are the best-known methods.

Single linkage clustering computes the (dis)similarity between two groups as the (dis)similarity between the closest pair of observations between the two groups. Complete linkage clustering, on the other hand, uses the farthest pair of observations between the two groups to determine the (dis)similarity of the two groups. Average linkage clustering uses the average (dis)similarity of observations between the groups as the measure between the two groups. Ward's method joins the two groups that result in the minimum increase in the error sum of squares. The other linkage methods provide alternatives to these basic linkage methods.

The `cluster singlelinkage` command implements single linkage hierarchical agglomerative clustering; see [CL] **cluster singlelinkage** for details. Single linkage clustering suffers (or benefits, depending on your point of view) from what is called chaining. Since the closest points between two groups determine the next merger, long, thin clusters can result. If this chaining feature is not what you desire, then consider using one of the other methods, such as complete linkage or average linkage. Single linkage clustering is faster and uses less memory than the other linkage methods due to special properties of the method that can be exploited computationally.

Complete linkage hierarchical agglomerative clustering is implemented by the `cluster completelinkage` command; see [CL] **cluster completelinkage** for details. Complete linkage clustering is at the other extreme from single linkage clustering. Complete linkage produces spatially compact clusters. Complete linkage clustering is not the best method for recovering elongated cluster structures. Several sources, including Kaufman and Rousseeuw (1990), discuss the chaining of single linkage and the clumping of complete linkage.

Average linkage hierarchical agglomerative cluster analysis has properties that are intermediate of single and complete linkage clustering. Kaufman and Rousseeuw (1990) indicate that average linkage works well for many situations and is reasonably robust. The `cluster averagelinkage` command provides average linkage clustering; see [CL] **cluster averagelinkage**.

Ward (1963) presented a general hierarchical clustering approach where groups were joined so as to maximize an objective function. He used an error-sum-of-squares objective function to illustrate. Ward's method of clustering became synonymous with using the error-sum-of-squares criteria. Kaufman and Rousseeuw (1990) indicate that Ward's method does well with groups that are multivariate normal and spherical, but does not do as well if the groups are of different size or have unequal numbers of observations. The `cluster wardslinkage` command provides Ward's linkage clustering; see [CL] **cluster wardslinkage**.

At each step of the clustering, centroid linkage merges the groups whose means are closest. The centroid of a group is the component-wise mean, and can be interpreted as the center of gravity for the group. Centroid linkage differs from average linkage in that centroid linkage is concerned with the distance between the means of the groups, while average linkage is looking at the average distance between the points of the two groups. The `cluster centroidlinkage` command provides centroid linkage clustering; see [CL] **cluster centroidlinkage**.

Weighted-average linkage and median linkage are variations on average linkage and centroid linkage, respectively. In both cases, the difference is in how groups of unequal size are treated when merged. In average linkage and centroid linkage, the number of elements of each group are factored into the computation, giving correspondingly larger influence to the larger group. These two methods are called unweighted because each observation carries the same weight. In weighted-average linkage and median linkage, the two groups are given equal weighting, regardless of the number of observations in each group, in determining the combined group. These two methods are said to be weighted since observations from groups with few observations carry more weight than observations from groups with many observations. The `cluster waveragelinkage` command provides weighted-average linkage clustering; see [CL] **cluster waveragelinkage**. The `cluster medianlinkage` command provides median linkage clustering; see [CL] **cluster medianlinkage**.

Lance and Williams' recurrence formula

Lance and Williams (1967) developed a recurrence formula that defines, as special cases, most of the well-known hierarchical clustering methods, including all of the hierarchical clustering methods found in Stata. Anderberg (1973), Jain and Dubes (1988), Kaufman and Rousseeuw (1990), Gordon (1999), Everitt, Landau, and Leese (2001), and Rencher (2002) discuss the Lance and Williams formula and how most popular hierarchical clustering methods are contained within it.

Using the notation of Everitt, Landau, and Leese (2001, 61), the Lance and Williams recurrence formula is

$$d_{k(ij)} = \alpha_i d_{ki} + \alpha_j d_{kj} + \beta d_{ij} + \gamma |d_{ki} - d_{kj}|$$

where d_{ij} is the distance (or dissimilarity) between cluster i and cluster j; $d_{k(ij)}$ is the distance (or dissimilarity) between cluster k and the new cluster formed by joining clusters i and j; and α_i, α_j, β, and γ are parameters that are set based on the particular hierarchical cluster analysis method.

The recurrence formula allows, at each new level of the hierarchical clustering, the dissimilarity between the newly formed group and the rest of the groups to be computed from the dissimilarities of the current grouping. This can result in a large computational savings compared with recomputing at each step in the hierarchy from the observation-level data.

The following table shows the values of α_i, α_j, β, and γ for the hierarchical clustering methods implemented in Stata. n_i, n_j, and n_k are the number of observations in group i, j, and k, respectively.

Clustering method	α_i	α_j	β	γ
Single linkage	$\frac{1}{2}$	$\frac{1}{2}$	0	$-\frac{1}{2}$
Complete linkage	$\frac{1}{2}$	$\frac{1}{2}$	0	$\frac{1}{2}$
Average linkage	$\frac{n_i}{n_i + n_j}$	$\frac{n_j}{n_i + n_j}$	0	0
Weighted-average linkage	$\frac{1}{2}$	$\frac{1}{2}$	0	0
Centroid linkage	$\frac{n_i}{n_i + n_j}$	$\frac{n_j}{n_i + n_j}$	$-\alpha_i\alpha_j$	0
Median linkage	$\frac{1}{2}$	$\frac{1}{2}$	$-\frac{1}{4}$	0
Ward's linkage	$\frac{n_i + n_k}{n_i + n_j + n_k}$	$\frac{n_j + n_k}{n_i + n_j + n_k}$	$\frac{-n_k}{n_i + n_j + n_k}$	0

For information on the use of various similarity and dissimilarity measures in hierarchical clustering, see the discussion in *(Dis)similarity transformations and the Lance and Williams formula* and in *Warning concerning (dis)similarity choice* below.

(Dis)similarity transformations and the Lance and Williams formula

The Lance and Williams formula, which is used as the basis for computing hierarchical clustering in Stata, is designed for use with dissimilarity measures. Similarity measures, both continuous and binary, are transformed by Stata to dissimilarities before performing the hierarchical clustering. The fusion values (heights at which the various groups join in the hierarchy) are transformed back to similarities at the completion of the cluster analysis.

Stata uses

$$\text{dissimilarity} = 1 - \text{similarity}$$

to transform from a similarity to a dissimilarity measure and back again. Stata's similarity measures range from either 0 to 1, or -1 to 1. The resulting dissimilarities range from 1 down to 0, and from 2 down to 0, respectively.

For continuous data, Stata provides both the L2 and L2squared dissimilarity measures, as well as both the L(#) and Lpower(#) dissimilarity measures. Why have both a L2 and L2squared dissimilarity measure, and why have both a L(#) and Lpower(#) dissimilarity measure?

For single and complete linkage hierarchical clustering (and for kmeans and kmedians partition clustering), there is no need for the additional L2squared and Lpower(#) dissimilarities. The same cluster solution is obtained using L2 and L2squared (or L(#) and Lpower(#)), other than the resulting heights in the dendrogram being raised to a power.

However, for the other hierarchical clustering methods, there is a difference, and hence, the need for these additional dissimilarity options. For some of these other hierarchical clustering methods, the natural default for dissimilarity measure is L2squared. For instance, the traditional Ward's (1963) method is obtained using the L2squared dissimilarity option.

Warning concerning (dis)similarity choice

With hierarchical centroid, median, Ward's, and weighted-average linkage clustering, Lance and Williams (1967), Anderberg (1973), Jain and Dubes (1988), Kaufman and Rousseeuw (1990), Everitt, Landau, and Leese (2001), and Gordon (1999) give various levels of warning concerning the use of many of the similarity and dissimilarity measures. These warnings range from saying that you should never use anything other than the default squared Euclidean distance (or Euclidean distance), to saying that the results may lack a useful interpretation.

The second example in [CL] **cluster wardslinkage** illustrates part of the basis for this warning. The simple matching coefficient is used on binary data. The range of the fusion values for the resulting hierarchy is not between 1 and 0 as you would expect for the matching coefficient. The conclusions from the cluster analysis, however, agrees well with the results obtained in other ways.

Stata leaves the choice of (dis)similarity and the interpretation of the resulting clustering to the user. If you are not familiar with the details of these hierarchical clustering methods, you are advised to use the default dissimilarity measure.

Synonyms

There are many names attached to each of the cluster analysis methods that have been developed. Cluster analysis methods have been developed by researchers in many different disciplines. Since researchers did not always know of the similar developments happening in other fields, many synonyms for the different hierarchical cluster analysis methods exist.

Blashfield and Aldenderfer (1978) provide a table of equivalent terms. Jain and Dubes (1988) and Day and Edelsbrunner (1984) also mention some of the synonyms and make use of various acronyms. Here is a list of synonyms:

(Continued on next page)

Single linkage
    Nearest-neighbor method
    Minimum method
    Hierarchical analysis
    Space-contracting method
    Elementary linkage analysis
    Connectedness method

Complete linkage
    Furthest-neighbor method
    Maximum method
    Compact method
    Space-distorting method
    Space-dilating method
    Rank order typal analysis
    Diameter analysis

Average linkage
    Arithmetic average clustering
    Unweighted pair-group method using
        arithmetic averages
    UPGMA
    Unweighted clustering
    Group-average method
    Unweighted group mean
    Unweighted pair-group method

Weighted-average linkage
    Weighted pair-group method using
        arithmetic averages
    WPGMA
    Weighted group average method

Centroid linkage
    Unweighted centroid method
    Unweighted pair-group centroid method
    UPGMC
    Nearest-centroid sorting

Median linkage
    Gower's method
    Weighted centroid method
    Weighted pair-group centroid method
    WPGMC
    Weighted pair method
    Weighted group method

Ward's method
    Minimum-variance method
    Error-sum-of-squares method
    Hierarchical grouping to minimize tr(W)
    HGROUP

Reversals

Unlike the other hierarchical methods implemented in Stata, centroid linkage ([CL] **cluster centroidlinkage**) and median linkage ([CL] **cluster medianlinkage**) can (and often do) produce reversals or crossovers. This is discussed by Anderberg (1973), Jain and Dubes (1988), Gordon (1999), and Rencher (2002). Normally, the dissimilarity or clustering criterion increases monotonically as the agglomerative hierarchical clustering progresses from many to few clusters. (For similarity measures, it monotonically decreases.) In other words, the dissimilarity value at which $k + 1$ clusters form will be larger than the value at which k clusters form. When the dissimilarity does not increase monotonically through the levels of the hierarchy, it is said to have reversals or crossovers.

The word crossover, in this context, comes from the appearance of the resulting dendrogram (see [CL] **cluster dendrogram**). In a hierarchical clustering without reversals, the dendrogram branches extend in one direction (increasing dissimilarity measure). With reversals, some of the branches reverse and go in the opposite direction, causing the resulting dendrogram to be drawn with crossing lines (crossovers).

When reversals happen, Stata still produces correct results. You can still generate grouping variables (see [CL] **cluster generate**) and compute stopping rules (see [CL] **cluster stop**). However, the cluster dendrogram command will not draw a dendrogram with reversals; see [CL] **cluster dendrogram**. In all but the simplest cases, dendrograms with reversals are almost impossible to visually interpret.

Post-clustering commands

Stata's `cluster stop` command is used to determine the number of clusters. Two stopping rules are provided, the Caliński & Harabasz (1974) pseudo-F index and the Duda & Hart (1973) Je(2)/Je(1) index with associated pseudo T-squared. Additional stopping rules can easily be added to the `cluster stop` command. See [CL] **cluster stop** for details.

The `cluster dendrogram` command presents the dendrogram (cluster tree) after a hierarchical cluster analysis; see [CL] **cluster dendrogram**. Options allow you to view the top portion of the tree or the portion of the tree associated with a group. These options are important with larger datasets, since the full dendrogram cannot be presented.

The `cluster generate` command produces grouping variables after hierarchical clustering; see [CL] **cluster generate**. These variables can then be used in other Stata commands, such as those that tabulate, summarize, and provide graphs. For instance, you might use `cluster generate` to create a grouping variable. You then might use the `pca` and `score` commands (see [R] **pca**) to obtain the first two principal components of the data, and follow that with a graph (see *Stata Graphics Reference Manual*) to plot the principal components, using the grouping variable from the `cluster generate` command to control the point labeling of the graph. This would allow you to get one type of view into the clustering behavior of your data.

Cluster management tools

You may add notes to your cluster analysis with the `cluster notes` command; see [CL] **cluster notes**. This command also allows you to view and to delete notes attached to the cluster analysis.

The `cluster dir` and `cluster list` commands allow you to list the cluster objects and attributes currently defined for your dataset. `cluster drop` lets you remove a cluster object. See [CL] **cluster utility** for details.

Cluster objects are referenced by name. Many of the `cluster` commands will, by default, use the cluster object from the most recently performed cluster analysis if no name is provided. The `cluster use` command tells Stata to set a particular cluster object as the latest. The name attached to a cluster object may be changed with the `cluster rename` command, and the variables associated with a cluster analysis may be renamed with the `cluster renamevar` command. See [CL] **cluster utility** for details.

Programmers, and regular users if they desire, can exercise fine control over the attributes that are stored with a cluster object; see [CL] **cluster programming utilities**.

References

Anderberg, M. R. 1973. *Cluster Analysis for Applications*. New York: Academic Press.

Blashfield, R. K. and M. S. Aldenderfer. 1978. The literature on cluster analysis *Multivariate Behavioral Research* 13: 271–295.

Caliński, T. and J. Harabasz. 1974. A dendrite method for cluster analysis. *Communications in Statistics* 3: 1–27.

Czekanowski, J. 1932. "Coefficient of racial likeness" und "durchschnittliche Differenz". *Anthropologischer Anzeiger* 9: 227–249.

Day, W. H. E. and H. Edelsbrunner. 1984. Efficient algorithms for agglomerative hierarchical clustering methods. *Journal of Classification* 1: 7–24.

Dice, L. R. 1945. Measures of the amount of ecologic association between species. *Ecology* 26: 297–302.

Duda, R. O. and P. E. Hart. 1973. *Pattern Classification and Scene Analysis*. New York: John Wiley & Sons.

Everitt, B. S. 1993. *Cluster Analysis.* 3d ed. London: Edward Arnold.

Everitt, B. S., S. Landau, and M. Leese. 2001. *Cluster Analysis.* 4th ed. London: Edward Arnold.

Gordon, A. D. 1999. *Classification.* 2d ed. Boca Raton, FL: CRC Press.

Gower, J. C. 1985. Measures of similarity, dissimilarity, and distance. In *Encyclopedia of Statistical Sciences*, Vol. 5, ed. S. Kotz, N. L. Johnson, and C. B. Read, 397–405. New York: John Wiley & Sons.

Guilford, J. P. 1942. *Fundamental Statistics in Psychology and Education.* New York: McGraw-Hill.

Hamman, U. 1961. Merkmalsbestand und Verwandtschaftsbeziehungen der Farinosae. Ein Beitrag zum System der Monokotyledonen. *Willdenowia* 2: 639–768.

Hilbe, J. 1992a. sg9: Similarity coefficients for 2 x 2 binary data. *Stata Technical Bulletin* 9: 14–15. Reprinted in *Stata Technical Bulletin Reprints*, vol. 2, pp. 130–131.

———. 1992b. sg9.1: Additional statistics to similari output. *Stata Technical Bulletin* 10: 22. Reprinted in *Stata Technical Bulletin Reprints*, vol. 2, p. 132.

Jaccard, P. 1908. Nouvelles recherches sur la distribution florale. *Bulletin de la Société Vaudoise des Sciences Naturelles* 44: 223–270.

Jain, A. K. and R. C. Dubes. 1988. *Algorithms for Clustering Data.* Englewood Cliffs, NJ: Prentice–Hall.

Kaufman, L. and P. J. Rousseeuw. 1990. *Finding Groups in Data.* New York: John Wiley & Sons.

Kulczynski, S. 1927. Die Pflanzenassoziationen der Pieninen. [In Polish, German summary.] *Bulletin International de l'Academie Polonaise des Sciences et des Lettres, Classe des Sciences Mathematiques et Naturelles, B (Sciences Naturelles)* 1927 (Suppl. 2): 57–203.

Lance, G. N. and W. T. Williams. 1967. A general theory of classificatory sorting strategies: 1. Hierarchical systems. *Computer Journal* 9: 373–380.

Milligan, G. W. and M. C. Cooper. 1985. An examination of procedures for determining the number of clusters in a dataset. *Psychometrika* 50: 159–179.

———. 1988. A study of standardization of variables in cluster analysis. *Journal of Classification* 5: 181–204.

Ochiai, A. 1957. Zoogeographic studies on the soleoid fishes found in Japan and its neighbouring regions. [In Japanese, English summary.] *Bulletin of the Japanese Society of Scientific Fisheries* 22: 526–530.

Rencher, A. C. 2002. *Methods of Multivariate Analysis.* 2d ed. New York: John Wiley & Sons.

Rogers, D. J. and T. T. Tanimoto. 1960. A computer program for classifying plants. *Science* 132: 1115–1118.

Rohlf, F. J. 1982. Single-link clustering algorithms. In *Handbook of Statistics*, Vol. 2, ed. P. R. Krishnaiah and L. N. Kanal, 267–284. Amsterdam: North–Holland Publishing Company.

Russell, P. F. and T. R. Rao. 1940. On habitat and association of species of anopheline larvae in south-eastern Madras. *Journal of the Malaria Institute of India* 3: 153–178.

Schaffer, C. M. and P. E. Green. 1996. An empirical comparison of variable standardization methods in cluster analysis. *Multivariate Behavioral Research* 31: 149–167.

Sibson, R. 1973. SLINK: An optimally efficient algorithm for the single-link cluster method. *Computer Journal* 16: 30–34.

Sneath, P. H. A. and R. R. Sokal. 1962. Numerical taxonomy. *Nature* 193: 855–860.

Sokal, R. R. and C. D. Michener. 1958. A statistical method for evaluating systematic relationships. *University of Kansas Science Bulletin* 38: 1409–1438.

Sørensen, T. 1948. A method of establishing groups of equal amplitude in plant sociology based on similarity of species content and its application to analyses of the vegetation on Danish commons. *Kongel. Danske Vidensk. Selsk. Biol. Skr.* 5(4): 1–34.

Späth, H. 1980. *Cluster Analysis Algorithms for Data Reduction and Classification of Objects.* Chichester, England: Ellis Horwood.

Ward, J. H., Jr. 1963. Hierarchical grouping to optimize an objective function. *Journal of the American Statistical Association* 58: 236–244.

Yule, G. U. and M. G. Kendall. 1950. *An Introduction to the Theory of Statistics.* 14th ed. New York: Hafner.

Also See

Complementary:	[CL] **cluster averagelinkage**, [CL] **cluster centroidlinkage**, [CL] **cluster completelinkage**, [CL] **cluster dendrogram**, [CL] **cluster generate**, [CL] **cluster kmeans**, [CL] **cluster kmedians**, [CL] **cluster medianlinkage**, [CL] **cluster notes**, [CL] **cluster singlelinkage**, [CL] **cluster stop**, [CL] **cluster utility**, [CL] **cluster wardslinkage**, [CL] **cluster waveragelinkage**
Related:	[CL] **cluster programming subroutines**, [CL] **cluster programming utilities**

Title

cluster averagelinkage — Average linkage cluster analysis

Syntax

cluster <u>a</u>veragelinkage [*varlist*] [if *exp*] [in *range*] [, <u>name</u>(*clname*)

    distance_option <u>gen</u>erate(*stub*)]

Description

The cluster averagelinkage command performs hierarchical agglomerative average linkage cluster analysis. See [CL] **cluster** for a general discussion of cluster analysis and for a description of the other cluster commands. The cluster dendrogram command (see [CL] **cluster dendrogram**) will display the resulting dendrogram, the cluster stop command (see [CL] **cluster stop**) will help in determining the number of groups, and the cluster generate command (see [CL] **cluster generate**) will produce grouping variables.

Options

name(*clname*) specifies the name to attach to the resulting cluster analysis. If name() is not specified, Stata finds an available cluster name, displays it for your reference, and then attaches the name to your cluster analysis.

distance_option is one of the similarity or dissimilarity measures allowed by Stata. Capitalization of the option does not matter. See [CL] **cluster** for a discussion of these measures.

The available measures designed for continuous data are L2 (synonym <u>Euc</u>lidean), which is the default; L2squared; L1 (synonyms <u>abs</u>olute, <u>city</u>block, and <u>man</u>hattan); <u>Linf</u>inity (synonym <u>max</u>imum); L(#); <u>Lp</u>ower(#); <u>Can</u>berra; <u>corr</u>elation; and <u>ang</u>ular (synonym angle).

The available measures designed for binary data are <u>mat</u>ching, <u>Jac</u>card, <u>Russ</u>ell, Hamman, Dice, antiDice, Sneath, Rogers, Ochiai, Yule, <u>Ander</u>berg, <u>Kulc</u>zynski, Gower2, and Pearson.

generate(*stub*) provides a prefix for the variable names created by cluster averagelinkage. By default, the variable-name prefix will be the name specified in name(). Three variables are created and attached to the cluster analysis results, with the suffixes _id, _ord, and _hgt. Users generally will not need to access these variables directly.

Remarks

An example using the default L2 (Euclidean) distance on continuous data and an example using the matching coefficient on binary data illustrate the cluster averagelinkage command. These are the same datasets used as examples in [CL] **cluster centroidlinkage**, [CL] **cluster completelinkage**, [CL] **cluster medianlinkage**, [CL] **cluster singlelinkage**, [CL] **cluster wardslinkage**, and [CL] **cluster waveragelinkage**, so that you can compare the results from using different hierarchical clustering methods.

▷ Example

As explained in the first example of [CL] **cluster singlelinkage**, as the senior data analyst for a small biotechnology firm, you are given a dataset with 4 chemical laboratory measurements on 50 different samples of a particular plant gathered from the rain forest. The head of the expedition that gathered the samples thinks, based on information from the natives, that an extract from the plant might reduce the negative side effects associated with your company's best-selling nutritional supplement.

While the company chemists and botanists continue exploring the possible uses of the plant and plan future experiments, the head of product development asks you to look at the preliminary data and to report anything that might be helpful to the researchers.

While all 50 of the plants are supposed to be of the same type, you decide to perform a cluster analysis to see if there are subgroups or anomalies among them. Single linkage clustering helped you discover an anomaly in the data. You now wish to see if you discover the same thing using average linkage clustering with the default Euclidean distance.

You first call cluster averagelinkage and use the name() option to attach the name L2alnk to the resulting cluster analysis. The cluster list command (see [CL] **cluster utility**) is then applied to list the components of your cluster analysis. The cluster dendrogram command then graphs the dendrogram; see [CL] **cluster dendrogram**. As described in the [CL] **cluster singlelinkage** example, the labels() option is used, instead of the default action of showing the observation number, to identify which laboratory technician produced the data.

```
. use http://www.stata-press.com/data/r8/labtech
. cluster averagelinkage x1 x2 x3 x4, name(L2alnk)
. cluster list L2alnk
L2alnk  (type: hierarchical,  method: average,  dissimilarity: L2)
       vars: L2alnk_id (id variable)
             L2alnk_ord (order variable)
             L2alnk_hgt (height variable)
      other: range: 0 .
             cmd: cluster averagelinkage x1 x2 x3 x4, name(L2alnk)
             varlist: x1 x2 x3 x4
. cluster dendrogram L2alnk, vertlab ylab labels(labtech)
```

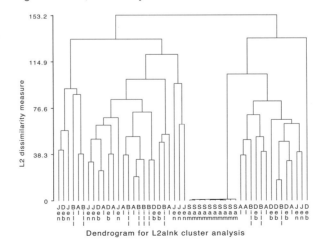

Dendrogram for L2alnk cluster analysis

As with single linkage clustering, you see that the samples analyzed by Sam, the lab technician, cluster together closely (dissimilarity measures near zero) and are separated from the rest of the data by a large dissimilarity gap (the long vertical line going up from Sam's cluster to eventually combine with other observations). When you examined the data, you discovered that Sam's data are all between zero and one, while the other four technicians have data that range from zero up to near 150. It appears that Sam has made a mistake.

If you compare the dendrogram from this average linkage clustering with those from single linkage clustering and complete linkage clustering, you will notice that the y-axis range is intermediate of these two other methods. This is a property of these linkage methods. With average linkage, it is the average of the (dis)similarities between the two groups that determines the distance between the groups. This is in contrast to the smallest distance and largest distance that define single linkage and complete linkage clustering.

◁

▷ Example

This example analyzes the same data as introduced in the second example of [CL] **cluster singlelinkage**. The sociology professor of your graduate-level class gives, as homework, a dataset containing 30 observations on 60 binary variables, with the assignment to tell him something about the 30 subjects represented by the observations.

In addition to examining single linkage clustering of these data, you decide to see what average linkage clustering shows. As with the single linkage clustering, you pick the simple matching binary coefficient to measure the similarity between groups. The `name()` option is used to attach the name `alink` to the cluster analysis. `cluster list` displays the details; see [CL] **cluster utility**. `cluster tree`, which is a synonym for `cluster dendrogram`, then displays the cluster tree (dendrogram); see [CL] **cluster dendrogram**.

(Continued on next page)

```
. use http://www.stata-press.com/data/r8/homework, clear
. cluster a a1-a60, match name(alink)
. cluster list alink
alink  (type: hierarchical,  method: average,  similarity: matching)
      vars: alink_id (id variable)
            alink_ord (order variable)
            alink_hgt (height variable)
     other: range: 1 0
            cmd: cluster averagelinkage a1-a60, match name(alink)
            varlist: a1 a2 a3 a4 a5 a6 a7 a8 a9 a10 a11 a12 a13 a14 a15 a16 a17
                a18 a19 a20 a21 a22 a23 a24 a25 a26 a27 a28 a29 a30 a31 a32
                a33 a34 a35 a36 a37 a38 a39 a40 a41 a42 a43 a44 a45 a46 a47
                a48 a49 a50 a51 a52 a53 a54 a55 a56 a57 a58 a59 a60
```

```
. cluster tree
```

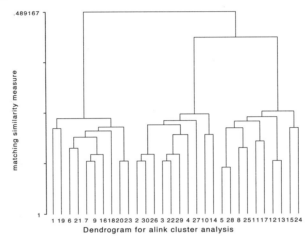

Dendrogram for alink cluster analysis

Since, by default, Stata uses the most recently performed cluster analysis, you do not need to type the cluster name when calling `cluster tree`.

As with single linkage clustering, the dendrogram from average linkage clustering seems to indicate the presence of 3 groups among the 30 observations. Later you receive another variable called `truegrp` that identifies the groups that the teacher believes are in the data. You use the `cluster generate` command (see [CL] **cluster generate**) to create a grouping variable, based on your average linkage clustering, to compare with `truegrp`. You do a cross-tabulation of `truegrp` and `agrp3`, your grouping variable, to see if your conclusions match those of the teacher.

```
. cluster gen agrp3 = group(3)
. table agrp3 truegrp
```

agrp3	truegrp 1	2	3
1		10	
2	10		
3			10

Other than the numbers arbitrarily assigned to the three groups, your teacher's conclusions and the results from the average linkage clustering are in complete agreement.

◁

❑ Technical Note

cluster averagelinkage requires more memory and more execution time than cluster singlelinkage. With a large number of observations, the execution time may be significant.

❑

Methods and Formulas

[CL] **cluster** discusses hierarchical clustering, and places average linkage clustering in this general framework. It compares the typical behavior of average linkage with that of single linkage and that of complete linkage.

Conceptually, hierarchical agglomerative average linkage clustering proceeds as follows. The N observations start out as N separate groups, each of size one. The two closest observations are merged into one group, producing $N - 1$ total groups. The closest two groups are then merged, so that there are $N - 2$ total groups. This process continues until all the observations are merged into one large group. This produces a hierarchy of groupings from one group to N groups. For average linkage clustering, the decision of "closest two groups" is based on the average (dis)similarity between the observations of the two groups.

The average linkage clustering algorithm produces two variables that act as a pointer representation of a dendrogram. To this, Stata adds a third variable used to restore the sort order, as needed, so that the two variables of the pointer representation remain valid. The first variable of the pointer representation gives the order of the observations. The second variable has one less element, and gives the height in the dendrogram at which the adjacent observations in the order-variable join.

See [CL] **cluster** for the details and formulas of the available *distance_options*, which include (dis)similarity measures for continuous and for binary data.

Also See

Complementary:	[CL] **cluster dendrogram**, [CL] **cluster generate**, [CL] **cluster notes**, [CL] **cluster stop**, [CL] **cluster utility**
Related:	[CL] **cluster centroidlinkage**, [CL] **cluster completelinkage**, [CL] **cluster medianlinkage**, [CL] **cluster singlelinkage**, [CL] **cluster wardslinkage**, [CL] **cluster waveragelinkage**
Background:	[CL] **cluster**

Title

cluster centroidlinkage — Centroid linkage cluster analysis

Syntax

cluster <u>centr</u>oidlinkage [*varlist*] [if *exp*] [in *range*] [, <u>n</u>ame(*clname*)

    distance_option <u>gene</u>rate(*stub*)]

Description

The cluster centroidlinkage command performs hierarchical agglomerative centroid linkage cluster analysis. See [CL] **cluster** for a general discussion of cluster analysis and for a description of the other cluster commands. The cluster dendrogram command (see [CL] **cluster dendrogram**) will display the resulting dendrogram, the cluster stop command (see [CL] **cluster stop**) will help in determining the number of groups, and the cluster generate command (see [CL] **cluster generate**) will produce grouping variables.

Options

name(*clname*) specifies the name to attach to the resulting cluster analysis. If name() is not specified, Stata finds an available cluster name, displays it for your reference, and then attaches the name to your cluster analysis.

distance_option is one of the similarity or dissimilarity measures allowed by Stata. Capitalization of the option does not matter. See [CL] **cluster** for a discussion of these measures.

The available measures designed for continuous data are L2 (synonym <u>Euclid</u>ean); L2squared, which is the default for cluster centroidlinkage; L1 (synonyms <u>absolute</u>, <u>city</u>block, and <u>manhattan</u>); <u>Linf</u>inity (synonym <u>maxi</u>mum); L(#); <u>Lpower</u>(#); <u>Canb</u>erra; <u>corr</u>elation; and angular (synonym <u>angle</u>).

The available measures designed for binary data are <u>match</u>ing, <u>Jacc</u>ard, <u>Russ</u>ell, Hamman, Dice, antiDice, Sneath, Rogers, Ochiai, Yule, <u>Ander</u>berg, <u>Kulc</u>zynski, Gower2, and Pearson.

Several authors advise the exclusive use of the L2squared *distance_option* with centroid linkage. See the sections *(Dis)similarity transformations and the Lance and Williams formula* and *Warning concerning (dis)similarity choice* in [CL] **cluster** for details.

generate(*stub*) provides a prefix for the variable names created by cluster centroidlinkage. By default, the variable-name prefix will be the name specified in name(). Three variables are created and attached to the cluster analysis results, with the suffixes _id, _ord, and _hgt. Users generally will not need to access these variables directly.

Centroid linkage can produce reversals or crossovers; see [CL] **cluster** for details. When reversals happen, cluster centroidlinkage also creates a fourth variable with the suffix _pht. This is a pseudo-height variable that is used by some of the post-clustering commands to properly interpret the _hgt variable.

Remarks

An example using the default L2squared (squared Euclidean) distance and L2 (Euclidean) distance on continuous data and an example using the matching coefficient on binary data illustrate the cluster centroidlinkage command. These are the same datasets introduced in [CL] **cluster singlelinkage**, which are used as examples for all the hierarchical clustering methods so that you can compare the results from using different hierarchical clustering methods.

▷ Example

As explained in the first example of [CL] **cluster singlelinkage**, as the senior data analyst for a small biotechnology firm, you are given a dataset with 4 chemical laboratory measurements on 50 different samples of a particular plant gathered from the rain forest. The head of the expedition that gathered the samples thinks, based on information from the natives, that an extract from the plant might reduce the negative side effects associated with your company's best-selling nutritional supplement.

While the company chemists and botanists continue exploring the possible uses of the plant and plan future experiments, the head of product development asks you to look at the preliminary data and to report anything that might be helpful to the researchers.

While all 50 of the plants are supposed to be of the same type, you decide to perform a cluster analysis to see if there are subgroups or anomalies among them. Single linkage clustering helped you discover an anomaly in the data. You now wish to see if you discover the same thing using centroid linkage clustering with the default squared Euclidean distance and with Euclidean distance.

You first call cluster centroidlinkage, letting the distance default to L2squared (squared Euclidean distance), and use the name() option to attach the name cent to the resulting cluster analysis. The cluster list command (see [CL] **cluster utility**) is then applied to list the components of your cluster analysis.

```
. use http://www.stata-press.com/data/r8/labtech

. cluster centroidlinkage x1 x2 x3 x4, name(cent)

. cluster list cent
cent  (type: hierarchical,  method: centroid,  dissimilarity: L2squared)
      vars: cent_id (id variable)
            cent_ord (order variable)
            cent_hgt (real_height variable)
            cent_pht (pseudo_height variable)
     other: range: 0 .
            cmd: cluster centroidlinkage x1 x2 x3 x4, name(cent)
            varlist: x1 x2 x3 x4
```

You do the same thing again, but this time using L2 (Euclidean distance) and giving it the name L2cent.

```
. cluster centroidlinkage x1 x2 x3 x4, name(L2cent) L2

. cluster list L2cent
L2cent  (type: hierarchical,  method: centroid,  dissimilarity: L2)
      vars: L2cent_id (id variable)
            L2cent_ord (order variable)
            L2cent_hgt (real_height variable)
            L2cent_pht (pseudo_height variable)
     other: range: 0 .
            cmd: cluster centroidlinkage x1 x2 x3 x4, name(L2cent) L2
            varlist: x1 x2 x3 x4
```

You wish to use the `cluster dendrogram` command to graph the dendrogram (see [CL] **cluster dendrogram**), but since this particular cluster analysis produces reversals, it refuses to draw the dendrogram.

You decide to use the `cluster generate` command (see [CL] **cluster generate**) to produce grouping variables for 2 to 10 groups for each of the two cluster analyses. You also wish to examine the cross-tabulation of each of these generated groups against a variable that identifies which laboratory technician produced the data. (For the sake of brevity, only one cross-tabulation is shown for each of the cluster analyses.)

```
. cluster gen gc = groups(2/10), name(cent)
. cluster gen gL2c = groups(2/10), name(L2cent)
. table labtech gc6
```

labtech	1	2	gc6 3	4	5	6
Al	1	5			4	
Bill	2	5			3	
Deb	1	4			4	1
Jen	2	3	3		2	
Sam				10		

```
. table labtech gL2c2
```

labtech	gL2c2 1	2
Al	10	
Bill	10	
Deb	10	
Jen	10	
Sam		10

The samples analyzed by Sam appear to want to stay clustered together more strongly than the samples analyzed by the other technicians. The reason for this phenomenon is not as obvious from this analysis as it was when viewing the dendrogram from the single linkage clustering (see [CL] **cluster singlelinkage**).

◁

▷ Example

This example analyzes the same data as introduced in the second example of [CL] **cluster singlelinkage**. The sociology professor of your graduate-level class gives, as homework, a dataset containing 30 observations on 60 binary variables, with the assignment to tell him something about the 30 subjects represented by the observations.

In addition to examining single linkage clustering of these data, you decide to see what centroid linkage clustering shows. As with the single linkage clustering, you pick the simple matching binary coefficient to measure the similarity between groups. The `name()` option is used to attach the name `centlink` to the cluster analysis. `cluster list` displays the details; see [CL] **cluster utility**.

```
. use http://www.stata-press.com/data/r8/homework
. cluster cent a1-a60, match name(centlink)
. cluster list centlink
centlink (type: hierarchical,  method: centroid,  similarity: matching)
      vars: centlink_id (id variable)
            centlink_ord (order variable)
            centlink_hgt (real_height variable)
            centlink_pht (pseudo_height variable)
     other: range: 1 0
            cmd: cluster centroidlinkage a1-a60, match name(centlink)
            varlist: a1 a2 a3 a4 a5 a6 a7 a8 a9 a10 a11 a12 a13 a14 a15 a16 a17
                     a18 a19 a20 a21 a22 a23 a24 a25 a26 a27 a28 a29 a30 a31 a32
                     a33 a34 a35 a36 a37 a38 a39 a40 a41 a42 a43 a44 a45 a46 a47
                     a48 a49 a50 a51 a52 a53 a54 a55 a56 a57 a58 a59 a60
```

You attempt to use the `cluster dendrogram` command to display the dendrogram, but since this particular cluster analysis produced reversals, `cluster dendrogram` refuses to produce the dendrogram. You realize that with reversals, the resulting dendrogram would not be easy to interpret anyway.

You decide to go directly to comparing the three-group solution from this centroid linkage clustering with the variable called `truegrp` provided by the teacher. You use the `cluster generate` command (see [CL] **cluster generate**) to create a grouping variable, based on your centroid clustering, to compare with `truegrp`.

```
. cluster gen centgrp3 = group(3)
. table centgrp3 truegrp
```

centgrp3	truegrp 1	2	3
1		10	
2	10		
3			10

Other than the numbers arbitrarily assigned to the three groups, your teacher's conclusions and the results from the three-group centroid linkage clustering are in agreement.

◁

❏ Technical Note

    `cluster centroidlinkage` requires more memory and more execution time than `cluster singlelinkage`. With a large number of observations, the execution time may be significant.

❏

Methods and Formulas

    [CL] **cluster** discusses hierarchical clustering, and places centroid linkage clustering in this general framework. Conceptually, hierarchical agglomerative clustering proceeds as follows. The N observations start out as N separate groups, each of size one. The two closest observations are merged into one group, producing $N - 1$ total groups. The closest two groups are then merged, so that there are $N - 2$ total groups. This process continues until all the observations are merged into one large group. This produces a hierarchy of groupings from one group to N groups. The difference between the various hierarchical linkage methods depends on how "closest" is defined when comparing groups. Centroid linkage merges the groups whose means are closest.

The centroid linkage clustering algorithm produces two variables that act as a pointer representation of a dendrogram. To this, Stata adds a third variable used to restore the sort order, as needed, so that the two variables of the pointer representation remain valid. The first variable of the pointer representation gives the order of the observations. The second variable has one less element, and gives the height in the dendrogram at which the adjacent observations in the order-variable join. When reversals happen, which they often do, a fourth variable, called a pseudo-height, is produced. This is used by post-clustering commands in conjunction with the height variable to properly interpret the ordering of the hierarchy.

See [CL] **cluster** for the details, warnings, and formulas of the available *distance_option*s, which include (dis)similarity measures for continuous and for binary data.

Also See

Complementary:	[CL] **cluster dendrogram**, [CL] **cluster generate**, [CL] **cluster notes**, [CL] **cluster stop**, [CL] **cluster utility**
Related:	[CL] **cluster averagelinkage**, [CL] **cluster completelinkage**, [CL] **cluster medianlinkage**, [CL] **cluster singlelinkage**, [CL] **cluster wardslinkage**, [CL] **cluster waveragelinkage**
Background:	[CL] **cluster**

Title

> **cluster completelinkage** — Complete linkage cluster analysis

Syntax

> cluster <u>c</u>ompletelinkage [*varlist*] [if *exp*] [in *range*] [, <u>na</u>me(*clname*)
>
> *distance_option* <u>gene</u>rate(*stub*)]

Description

The cluster completelinkage command performs hierarchical agglomerative complete linkage cluster analysis, which is also known (among other names) as the furthest-neighbor technique. See [CL] **cluster** for a general discussion of cluster analysis and for a description of the other cluster commands. The cluster dendrogram command (see [CL] **cluster dendrogram**) will display the resulting dendrogram, the cluster stop command (see [CL] **cluster stop**) will help in determining the number of groups, and the cluster generate command (see [CL] **cluster generate**) will produce grouping variables.

Options

name(*clname*) specifies the name to attach to the resulting cluster analysis. If name() is not specified, Stata finds an available cluster name, displays it for your reference, and then attaches the name to your cluster analysis.

distance_option is one of the similarity or dissimilarity measures allowed by Stata. Capitalization of the option does not matter. See [CL] **cluster** for a discussion of these measures.

The available measures designed for continuous data are L2 (synonym <u>Eucl</u>idean), which is the default; L2squared; L1 (synonyms <u>absolute</u>, <u>cityblock</u>, and <u>manhattan</u>); <u>Linf</u>inity (synonym <u>maximum</u>); L(#); <u>Lpow</u>er(#); <u>Can</u>berra; <u>corr</u>elation; and <u>ang</u>ular (synonym <u>ang</u>le).

The available measures designed for binary data are <u>match</u>ing, <u>Jac</u>card, <u>Russ</u>ell, Hamman, Dice, antiDice, Sneath, Rogers, Ochiai, Yule, <u>Ander</u>berg, <u>Kulc</u>zynski, Gower2, and Pearson.

generate(*stub*) provides a prefix for the variable names created by cluster completelinkage. By default, the variable-name prefix will be the name specified in name(). Three variables are created and attached to the cluster analysis results, with the suffixes _id, _ord, and _hgt. Users generally will not need to access these variables directly.

Remarks

An example using the default L2 (Euclidean) distance on continuous data and an example using the matching coefficient on binary data illustrate the cluster completelinkage command. These are the same datasets used as examples in [CL] **cluster averagelinkage**, [CL] **cluster centroidlinkage**, [CL] **cluster medianlinkage**, [CL] **cluster singlelinkage**, [CL] **cluster wardslinkage**, and [CL] **cluster waveragelinkage**, so that you can compare the results from using different hierarchical clustering methods.

▷ Example

As explained in the first example of [CL] **cluster singlelinkage**, as the senior data analyst for a small biotechnology firm, you are given a dataset with 4 chemical laboratory measurements on 50 different samples of a particular plant gathered from the rain forest. The head of the expedition that gathered the samples thinks, based on information from the natives, that an extract from the plant might reduce the negative side effects associated with your company's best-selling nutritional supplement.

While the company chemists and botanists continue exploring the possible uses of the plant and plan future experiments, the head of product development asks you to look at the preliminary data and to report anything that might be helpful to the researchers.

While all 50 of the plants are supposed to be of the same type, you decide to perform a cluster analysis to see if there are subgroups or anomalies among them. Single linkage clustering helped you discover an anomaly in the data. You now wish to see if you discover the same thing using complete linkage clustering with the default Euclidean distance.

You first call `cluster completelinkage` and use the `name()` option to attach the name L2clnk to the resulting cluster analysis. The `cluster list` command (see [CL] **cluster utility**) is then applied to list the components of your cluster analysis. The `cluster dendrogram` command then graphs the dendrogram; see [CL] **cluster dendrogram**. As described in the [CL] **cluster singlelinkage** example, the `labels()` option is used, instead of the default action of showing the observation number, to identify which laboratory technician produced the data.

```
. use http://www.stata-press.com/data/r8/labtech
. cluster completelinkage x1 x2 x3 x4, name(L2clnk)
. cluster list L2clnk
L2clnk  (type: hierarchical,  method: complete,  dissimilarity: L2)
        vars: L2clnk_id (id variable)
              L2clnk_ord (order variable)
              L2clnk_hgt (height variable)
       other: range: 0 .
              cmd: cluster completelinkage x1 x2 x3 x4, name(L2clnk)
              varlist: x1 x2 x3 x4
. cluster dendrogram L2clnk, vertlab ylab labels(labtech)
```

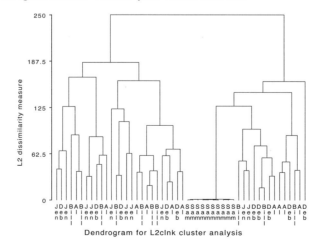

Dendrogram for L2clnk cluster analysis

As with single linkage clustering, you see that the samples analyzed by Sam, the lab technician, cluster together closely (dissimilarity measures near zero) and are separated from the rest of the data by a large dissimilarity gap (the long vertical line going up from Sam's cluster to eventually combine with other observations). When you examined the data, you discovered that Sam's data are all between zero and one, while the other four technicians have data that range from zero up to near 150. It appears that Sam has made a mistake.

If you compare the dendrogram from this complete linkage clustering with those from single linkage clustering and average linkage clustering, you will notice that the vertical lines at the top of the tree are relatively longer and the y-axis range is larger. This is a property of these linkage methods. The distance between groups is larger for complete linkage, since, by definition, with complete linkage the distance between two groups is the distance between their farthest members.

◁

> ◁ Example

This example analyzes the same data as introduced in the second example of [CL] **cluster singlelinkage**. The sociology professor of your graduate-level class gives, as homework, a dataset containing 30 observations on 60 binary variables, with the assignment to tell him something about the 30 subjects represented by the observations.

In addition to examining single linkage clustering of these data, you decide to see what complete linkage clustering shows. As with the single linkage clustering, you pick the simple matching binary coefficient to measure the similarity between groups. The name() option is used to attach the name clink to the cluster analysis. cluster list displays the details; see [CL] **cluster utility**. cluster tree, which is a synonym for cluster dendrogram, then displays the cluster tree (dendrogram); see [CL] **cluster dendrogram**.

(Continued on next page)

```
. use http://www.stata-press.com/data/r8/homework

. cluster c a1-a60, match name(clink)

. cluster list clink
clink  (type: hierarchical,  method: complete,  similarity: matching)
      vars: clink_id (id variable)
            clink_ord (order variable)
            clink_hgt (height variable)
     other: range: 1 0
            cmd: cluster completelinkage a1-a60, match name(clink)
            varlist: a1 a2 a3 a4 a5 a6 a7 a8 a9 a10 a11 a12 a13 a14 a15 a16 a17
                 a18 a19 a20 a21 a22 a23 a24 a25 a26 a27 a28 a29 a30 a31 a32
                 a33 a34 a35 a36 a37 a38 a39 a40 a41 a42 a43 a44 a45 a46 a47
                 a48 a49 a50 a51 a52 a53 a54 a55 a56 a57 a58 a59 a60

. cluster tree
```

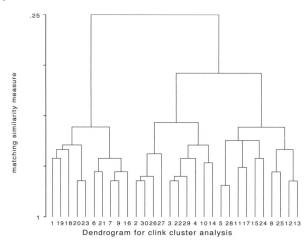

Dendrogram for clink cluster analysis

Since, by default, Stata uses the most recently performed cluster analysis, you do not need to type the cluster name when calling `cluster tree`.

As with single linkage clustering, the dendrogram from complete linkage clustering seems to indicate the presence of 3 groups among the 30 observations. Later you receive another variable called `truegrp` that identifies the groups that the teacher believes are in the data. You use the `cluster generate` command (see [CL] **cluster generate**) to create a grouping variable, based on your complete linkage clustering, to compare with `truegrp`. You do a cross-tabulation of `truegrp` and `cgrp3`, your grouping variable, to see if your conclusions match those of the teacher.

```
. cluster gen cgrp3 = group(3)

. table cgrp3 truegrp
```

	truegrp		
cgrp3	1	2	3
1		10	
2	10		
3			10

Other than the numbers arbitrarily assigned to the three groups, your teacher's conclusions and the results from the complete linkage clustering are in complete agreement.

◁

❑ Technical Note

> `cluster completelinkage` requires more memory and more execution time than `cluster singlelinkage`. With a large number of observations, the execution time may be significant.
>
> ❑

Methods and Formulas

[CL] **cluster** discusses hierarchical clustering, and places complete linkage clustering in this general framework. It compares the typical behavior of complete linkage with that of single linkage and that of average linkage.

Conceptually, hierarchical agglomerative complete linkage clustering proceeds as follows. The N observations start out as N separate groups each of size one. The two closest observations are merged into one group, producing $N - 1$ total groups. The closest two groups are then merged, so that there are $N - 2$ total groups. This process continues until all the observations are merged into one large group. This produces a hierarchy of groupings from one group to N groups. For complete linkage clustering, the definition of "closest two groups" is based on the farthest observations between the two groups.

The complete linkage clustering algorithm produces two variables that act as a pointer representation of a dendrogram. To this, Stata adds a third variable used to restore the sort order, as needed, so that the two variables of the pointer representation remain valid. The first variable of the pointer representation gives the order of the observations. The second variable has one less element, and gives the height in the dendrogram at which the adjacent observations in the order-variable join.

See [CL] **cluster** for the details and formulas of the available *distance_options*, which include (dis)similarity measures for continuous and for binary data.

Also See

Complementary:	[CL] **cluster dendrogram**, [CL] **cluster generate**, [CL] **cluster notes**, [CL] **cluster stop**, [CL] **cluster utility**
Related:	[CL] **cluster averagelinkage**, [CL] **cluster centroidlinkage**, [CL] **cluster medianlinkage**, [CL] **cluster singlelinkage**, [CL] **cluster wardslinkage**, [CL] **cluster waveragelinkage**
Background:	[CL] **cluster**

Title

| **cluster dendrogram** — Dendrograms for hierarchical cluster analysis |

Syntax

cluster <u>dend</u>rogram [*clname*] [if *exp*] [in *range*] [, quick <u>label</u>s(*varname*)

<u>vertl</u>abels <u>cutn</u>umber(#) <u>cutv</u>alue(#) <u>labcut</u>n <u>sa</u>ving(*filename*[, replace])

title_options axes_options]

Note: cluster <u>tree</u> is a synonym for cluster dendrogram.

In addition to the restrictions imposed by if and in, the observations are automatically restricted to those that were used in the formation of the cluster analysis.

Description

cluster dendrogram produces dendrograms (also called cluster trees) for a hierarchical clustering. See [CL] **cluster** for a discussion of cluster analysis, hierarchical clustering, and the available cluster commands.

Dendrograms graphically present the information concerning which observations are grouped together at various levels of (dis)similarity. At the bottom of the dendrogram, each observation is considered its own cluster. Vertical lines extend up for each observation, and at various (dis)similarity values these lines are connected to the lines from other observations with a horizontal line. The observations continue to combine until, at the top of the dendrogram, all observations are grouped together.

The height of the vertical lines and the range of the (dis)similarity axis give visual clues about the strength of the clustering. Long vertical lines indicate more distinct separation between the groups. Long vertical lines at the top of the dendrogram indicate that the groups represented by those lines are well separated from one another. Shorter lines indicate groups that are not as distinct.

Options

quick switches to a different style of dendrogram where the vertical lines only go straight up from the observations, instead of the default action of recentering the lines after each merge of observations in the dendrogram hierarchy. Some people prefer this representation, and it is quicker to render.

labels(*varname*) indicates that *varname* is to be used in place of observation numbers for labeling the observations at the bottom of the dendrogram.

vertlabels indicates that the labeling for the observations at the bottom of the dendrogram is to be presented vertically instead of horizontally. This is helpful when the labels have several characters and there are enough observations in the dendrogram to cause the labels to overlap.

cutnumber(#) displays only the top # branches of the dendrogram. With large dendrograms, the lower levels of the tree become too crowded. With cutnumber(), you can limit your view to the upper portion of the dendrogram. Also see the cutvalue() and labcutn options.

cutvalue(#) displays only the top portion of the dendrogram. Only those branches of the dendrogram that are above the # (dis)similarity measure are presented. With large dendrograms, the lower levels of the tree become too crowded. With cutvalue(), you can limit your view to the upper portion of the dendrogram. Also see the cutnumber() and labcutn options.

labcutn requests that the number of observations associated with each branch be displayed below the branches. labcutn is allowed only with cutnumber() or with cutvalue() since, otherwise, the number of observations for each branch is one.

saving(*filename*[, replace]) saves the graph in a file that can be reviewed by graph using (see *Stata Graphics Reference Manual*) and printed by pulling down the **File** menu and choosing **Print Graph**. If you do not specify an extension, .gph will be assumed.

Allowed *title_options* are
    title("*text*"),
    t1title("*text*"), t2title("*text*"),   b1title("*text*"), b2title("*text*"),
    l1title("*text*"), l2title("*text*"),   r1title("*text*"), and r2title("*text*").

Allowed *axes_options* are
    noaxis, gap(#),
    ylabel[(*numlist*)], rlabel[(*numlist*)],   ytick(*numlist*), and rtick(*numlist*).

Remarks

Examples of the cluster dendrogram command can be found in [CL] **cluster singlelinkage**, [CL] **cluster completelinkage**, [CL] **cluster averagelinkage**, [CL] **cluster wardslinkage**, [CL] **cluster waveragelinkage**, [CL] **cluster stop**, and [CL] **cluster generate**. Here we illustrate some of the additional options available with cluster dendrogram.

▷ Example

In the first example of [CL] **cluster completelinkage**, the dendrogram for the complete linkage clustering of 50 observations on 4 variables was illustrated with the following dendrogram:

    . use http://www.stata-press.com/data/r8/labtech
    . cluster completelinkage x1 x2 x3 x4, name(L2clnk)
    . cluster dendrogram L2clnk, vertlab ylab labels(labtech)

(*Continued on next page*)

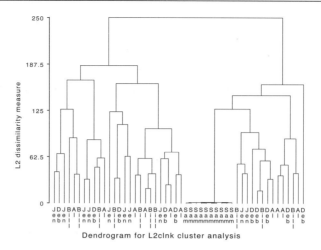

Dendrogram for L2clnk cluster analysis

This same dendrogram can be rendered in a slightly different format with the `quick` option:

```
. cluster dendrogram L2clnk, vertlab ylab labels(labtech) quick
```

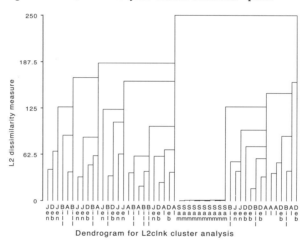

Dendrogram for L2clnk cluster analysis

Some people prefer this style of dendrogram. The option name `quick` comes from the fact that this style of dendrogram is quicker to render.

To display the dendrogram for one subgroup, use the `if` and `in` conditions to restrict to the subgroup observations. This is usually accomplished with the `cluster generate` command, which creates a grouping variable; see [CL] **cluster generate**.

Here we show the third of three groups in the dendrogram by first generating the grouping variable for three groups, and then using `if` in the command for `cluster dendrogram` to restrict to the third of those three groups.

```
. cluster gen g3 = group(3)
. cluster tree if g3==3, ylab
```

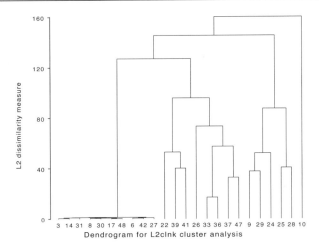

Dendrogram for L2clnk cluster analysis

Since we find it easier to type, we used the synonym `tree` instead of `dendrogram`. We did not specify the cluster name, instead allowing it to default to the most recently performed cluster analysis. We also omitted the `vertlabels` and `labels()` options, which bring us back to the default action of showing, horizontally, the observation numbers.

This example has only 50 observations. When there are a large number of observations, the dendrogram becomes too busy. You will need to limit which part of the dendrogram you display. One way to view a subpart of the dendrogram is to use `if` and `in` to limit to one particular group, as we did above.

The other way you can limit your view of the dendrogram is to specify that you only wish to view the top portion of the tree. The `cutnumber()` and `cutvalue()` options allow you to do this:

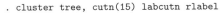

```
. cluster tree, cutn(15) labcutn rlabel
```

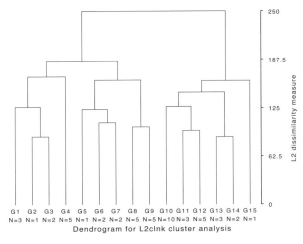

Dendrogram for L2clnk cluster analysis

We limited our view to the top 15 branches of the dendrogram with `cutn(15)`. By default, the 15 branches were labeled `G1` through `G15`. The `labcutn` option provided, below these branch labels, the number of observations in each of the 15 groups. And, just for variety, we specified the `rlabel` option to place the dissimilarity measure axis on the right instead of on the left.

The `cutvalue()` option provides another method of limiting the view to the top branches of the dendrogram. With this option, you specify the similarity or dissimilarity value at which to trim the tree.

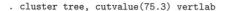

. `cluster tree, cutvalue(75.3) vertlab`

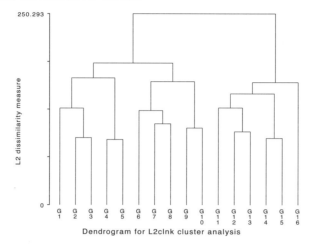

This time we limited the dendrogram to those branches with dissimilarity greater than 75.3 by using the `cutvalue(75.3)` option. There were 16 branches (groups) that met that restriction. We also specified the `vertlabels` option, but did not specify the `labcutn` option. This listed the 16 group labels vertically, and omitted the number of observations in each group.

The various title options allow you to add titles to the four sides of the dendrogram. The allowed axes options give you control over how, if at all, the (dis)similarity axis is to be displayed. You can experiment with these options to find the look that you like best.

◁

❏ Technical Note

Programmers can exercise control over what graphical procedure is executed when `cluster dendrogram` is called. This will be helpful to programmers adding new hierarchical clustering methods that require a different dendrogram algorithm. See [CL] **cluster programming subroutines** for details.

❏

Also See

Complementary:	[CL] **cluster averagelinkage**, [CL] **cluster centroidlinkage**, [CL] **cluster completelinkage**, [CL] **cluster generate**, [CL] **cluster medianlinkage**, [CL] **cluster programming subroutines**, [CL] **cluster singlelinkage**, [CL] **cluster stop**, [CL] **cluster wardslinkage**, [CL] **cluster waveragelinkage**
Background:	[CL] **cluster**

Title

cluster generate — Generate summary or grouping variables from a cluster analysis

Syntax

> cluster generate { *newvarname* | *stub* } = groups(*numlist*) [, name(*clname*)
>
>   ties(error | skip | less | more)]
>
> cluster generate *newvarname* = cut(*#*) [, name(*clname*)]

Description

The cluster generate command generates summary or grouping variables from a cluster analysis. What is produced depends on the function. See [CL] **cluster** for information on available cluster analysis commands.

The groups(*numlist*) function generates grouping variables, giving the grouping for the specified number(s) of clusters from a hierarchical cluster analysis. If a single number is given, *newvarname* is produced with group numbers going from 1 to the number of clusters requested. If more than one number is specified, a new variable is generated for each number using the provided *stub* name appended with the number. For instance,

```
cluster gen xyz = groups(5/7), name(myclus)
```

creates variables xyz5, xyz6, and xyz7, giving the five, six, and seven groups obtained from the cluster analysis named myclus.

The cut(*#*) function generates a grouping variable corresponding to cutting the dendrogram (see [CL] **cluster dendrogram**) of a hierarchical cluster analysis at the specified (dis)similarity value.

Additional cluster generate functions may be added; see [CL] **cluster programming subroutines**.

Options

name(*clname*) specifies the name of the cluster analysis to use in producing the new variables. The default is the last performed cluster analysis, which can be reset using the cluster use command; see [CL] **cluster utility**.

ties(error | skip | less | more) indicates what to do with the groups() function in the case of ties. A hierarchical cluster analysis has ties when multiple groups are generated at a particular (dis)similarity value. For example, you might have the case where you can uniquely create two, three, and four groups, but the next possible grouping produces eight groups due to ties.

ties(error), the default, produces an error message, and does not generate the requested variables.

ties(skip) indicates that the offending requests are to be ignored. No error message is produced, and only the requests that produce unique groupings will be honored. With multiple values specified in the groups() function, ties(skip) allows the processing of those that produce unique groupings, ignoring the rest.

41

`ties(less)` produces the results for the largest number of groups less than or equal to your request. In the example above with `groups(6)` and using `ties(less)`, you would get the same result as you would by using `groups(4)`.

`ties(more)` produces the results for the smallest number of groups greater than or equal to your request. In the example above with `groups(6)` and using `ties(more)`, you would get the same result as you would by using `groups(8)`.

Remarks

Examples of the use of the `groups()` function of `cluster generate` can be found in [CL] **cluster singlelinkage**, [CL] **cluster completelinkage**, [CL] **cluster averagelinkage**, [CL] **cluster centroidlinkage**, [CL] **cluster medianlinkage**, [CL] **cluster wardslinkage**, [CL] **cluster waveragelinkage**, [CL] **cluster stop**, and [CL] **cluster dendrogram**. Additional examples of the `groups()` and `cut()` functions of `cluster generate` are provided here.

Visually, these functions are best understood with a reference to the dendrogram from a hierarchical cluster analysis. The `cluster dendrogram` command produces dendrograms (cluster trees) from a hierarchical cluster analysis; see [CL] **cluster dendrogram**.

▷ Example

The first example of [CL] **cluster completelinkage** performs a complete linkage cluster analysis on 50 observations with 4 variables. The dendrogram presented there was labeled at the bottom by the name of the laboratory technician responsible for the data. Here we reproduce that same dendrogram, but use the default action of placing observation numbers as labels. We then use the `groups()` function of `cluster generate` to produce a grouping variable, splitting the data into two groups.

```
. use http://www.stata-press.com/data/r8/labtech
. cluster completelinkage x1 x2 x3 x4, name(L2clnk)
. cluster dendrogram L2clnk, vertlab ylab
```

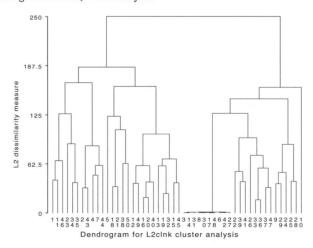

```
. cluster generate g2 = group(2), name(L2clnk)
```

```
. codebook g2
```

---

g2   (unlabeled)

```
                  type:  numeric (byte)
                 range:  [1,2]                      units:  1
         unique values:  2                     missing .:  0/50
            tabulation:  Freq.  Value
                           26   1
                           24   2
. bysort g2 : summarize x*
```

-> g2 = 1

Variable	Obs	Mean	Std. Dev.	Min	Max
x1	26	91.5	37.29432	17.4	143
x2	26	74.58077	41.19319	4.8	142.1
x3	26	101.0077	36.95704	16.3	147.9
x4	26	71.77308	43.04107	6.6	146.1

-> g2 = 2

Variable	Obs	Mean	Std. Dev.	Min	Max
x1	24	18.8	23.21742	0	77
x2	24	30.05833	37.66979	0	143.6
x3	24	18.54583	21.68215	.2	69.7
x4	24	41.89167	43.62025	.1	130.9

The group() function of cluster generate created a grouping variable named g2, with ones indicating the 26 observations belonging to the left main branch of the dendrogram, and twos indicating the 24 observations belonging to the right main branch of the dendrogram. The summary of the x variables used in the cluster analysis for each group shows that the second group is characterized by lower values.

We could have obtained the same grouping variable by using the cut() function of cluster generate.

```
. cluster gen g2cut = cut(200)
. table g2 g2cut
```

g2	g2cut 1	2
1	26	
2		24

Looking at the y-axis of the dendrogram, we decide to cut the tree at the dissimilarity value of 200. We did not specify the name() option. Instead, since this was the latest cluster analysis performed, we let it default to this latest cluster analysis. The table output shows that we obtained the same result with cut(200) as with group(2) for this example.

How many groups are produced if we cut the tree at the value 105.2?

```
. cluster gen z = cut(105.2)

. codebook z, tabulate(20)
```

---

z   (unlabeled)

```
         type:  numeric (byte)
        range:  [1,11]                        units:  1
unique values:  11                         missing .:  0/50
   tabulation:  Freq.  Value
                   3   1
                   3   2
                   5   3
                   1   4
                   2   5
                   2   6
                  10   7
                  10   8
                   8   9
                   5   10
                   1   11
```

The codebook command shows that the result of cutting the dendrogram at the value 105.2 produced eleven groups, ranging in size from one to ten observations.

The group() function of cluster generate may be used to create multiple grouping variables with a single call. Here we create the grouping variables for groups of size three to twelve:

```
. cluster gen gp = gr(3/12)

. summarize gp*
```

Variable	Obs	Mean	Std. Dev.	Min	Max
gp3	50	2.26	.8033095	1	3
gp4	50	3.14	1.030356	1	4
gp5	50	3.82	1.438395	1	5
gp6	50	3.84	1.461897	1	6
gp7	50	3.96	1.603058	1	7
gp8	50	4.24	1.911939	1	8
gp9	50	5.18	2.027263	1	9
gp10	50	5.94	2.385415	1	10
gp11	50	6.66	2.781939	1	11
gp12	50	7.24	3.197959	1	12

In this case, we used abbreviations for generate and group(). The group() function takes a numlist; see [U] **14.1.8 numlist**. We specified 3/12, which indicates the numbers 3 to 12. gp, the stub name we provide, is appended with the number as the variable name for each group variable produced.

◁

▷ Example

The second example of [CL] **cluster singlelinkage** shows the following dendrogram from the single linkage clustering of 30 observations on 60 variables. In that example, we used the group() function of cluster generate to produce a grouping variable for three groups. What happens when we try to obtain four groups from this clustering?

```
. use http://www.stata-press.com/data/r8/homework
. cluster singlelinkage a1-a60, matching
cluster name: _cl_1
. cluster tree
```

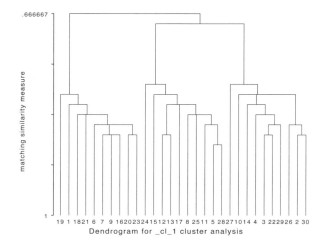

```
. cluster gen g4 = group(4)
cannot create 4 groups due to ties
r(198);
```

Stata complains that it cannot create four groups from this cluster analysis.

The ties() option gives us control over this situation. We just need to decide whether we want more groups or fewer groups than we asked for when faced with ties. We demonstrate both ways.

```
. cluster gen more4 = gr(4), ties(more)
. cluster gen less4 = gr(4), ties(less)
. summarize more4 less4
```

Variable	Obs	Mean	Std. Dev.	Min	Max
more4	30	2.933333	1.638614	1	5
less4	30	2	.8304548	1	3

For this cluster analysis, ties(more) with group(4) produces five groups, while ties(less) with group(4) produces three groups.

The ties(skip) option is convenient when we want to produce a range of grouping variables.

```
. cluster gen group = gr(4/20), ties(skip)
. summarize group*
```

Variable	Obs	Mean	Std. Dev.	Min	Max
group5	30	2.933333	1.638614	1	5
group9	30	4.866667	2.622625	1	9
group13	30	7.066667	3.92106	1	13
group18	30	9.933333	5.419844	1	18

With this cluster analysis, the only unique groupings available are 5, 9, 13, and 18 within the range 4 to 20.

◁

Also See

Complementary:	[CL] **cluster averagelinkage**, [CL] **cluster centroidlinkage**, [CL] **cluster completelinkage**, [CL] **cluster dendrogram**, [CL] **cluster medianlinkage**, [CL] **cluster programming subroutines**, [CL] **cluster singlelinkage**, [CL] **cluster stop**, [CL] **cluster wardslinkage**, [CL] **cluster waveragelinkage**
Related:	[R] **egen**, [R] **generate**
Background:	[CL] **cluster**

Title

> **cluster kmeans** — Kmeans cluster analysis

Syntax

> cluster <u>k</u>means [*varlist*] [if *exp*] [in *range*] , k(*#*) [<u>n</u>ame(*clname*)
>
>   *distance_option* <u>st</u>art(*start_option*) <u>gen</u>erate(*groupvar*) <u>iter</u>ate(*#*) <u>keep</u>centers]

Description

     cluster kmeans performs kmeans partition cluster analysis. See [CL] **cluster** for a general discussion of cluster analysis and for a description of the other cluster commands. See [CL] **cluster kmedians** for an alternative that uses medians instead of means.

Options

     k(*#*) is required, and indicates that *#* groups are to be formed by the cluster analysis.

     name(*clname*) specifies the name to attach to the resulting cluster analysis. If name() is not specified, Stata finds an available cluster name, displays it for your reference, and then attaches the name to your cluster analysis.

     distance_option is one of the similarity or dissimilarity measures allowed by Stata. Capitalization of the option does not matter. See [CL] **cluster** for a discussion of these measures.

     The available measures designed for continuous data are L2 (synonym <u>Euc</u>lidean), which is the default; L2squared; L1 (synonyms <u>abs</u>olute, <u>city</u>block, and <u>man</u>hattan); <u>Linf</u>inity (synonym <u>max</u>imum); L(*#*); <u>Lp</u>ower(*#*); <u>Can</u>berra; <u>corr</u>elation; and <u>ang</u>ular (synonym <u>angle</u>).

     The available measures designed for binary data are <u>matchi</u>ng, <u>Jac</u>card, <u>Russ</u>ell, Hamman, Dice, antiDice, Sneath, Rogers, Ochiai, Yule, <u>Ander</u>berg, <u>Kulc</u>zynski, Gower2, and Pearson.

     start(*start_option*) indicates how the *k* initial group centers are to be obtained. The available *start_option*s are <u>k</u>random[(*seed#*)], <u>f</u>irstk[, <u>ex</u>clude], <u>l</u>astk[, <u>ex</u>clude], <u>p</u>random[(*seed#*)], everykth, segments, group(*varname*), and <u>r</u>andom[(*seed#*)].

     krandom[(*seed#*)], the default, indicates that *k* unique observations are to be chosen at random, from among those to be clustered, as starting centers for the *k* groups. Optionally, a random number seed may be specified to cause the command set seed *seed#* (see [R] **generate**) to be applied before the *k* random observations are chosen.

     firstk[, exclude] indicates that the first *k* observations, from among those to be clustered, are to be used as the starting centers for the *k* groups. With the addition of the exclude option, these first *k* observations are then not included among the observations to be clustered.

     lastk[, exclude] indicates that the last *k* observations, from among those to be clustered, are to be used as the starting centers for the *k* groups. With the addition of the exclude option, these last *k* observations are then not included among the observations to be clustered.

prandom$\left[(seed\#)\right]$ indicates that k partitions are to be formed randomly among the observations to be clustered. The group means from the k groups defined by this partitioning are used as the starting group centers. Optionally, a random number seed may be specified to cause the command set seed *seed#* (see [R] **generate**) to be applied before the k partitions are chosen.

everykth indicates that k partitions are to be formed by assigning observations 1, $1 + k$, $1 + 2k$, . . . to the first group; assigning observations 2, $2 + k$, $2 + 2k$, . . . to the second group; and so on, to form k groups. The group means from these k groups are used as the starting group centers.

segments indicates that k nearly equal partitions are to be formed from the data. Approximately the first N/k observations are assigned to the first group, the second N/k observations are assigned to the second group, and so on. The group means from these k groups are used as the starting group centers.

group(*varname*) provides an initial grouping variable, *varname*, that defines k groups among the observations to be clustered. The group means from these k groups are used as the starting group centers.

random$\left[(seed\#)\right]$ indicates that k random initial group centers are to be generated. The values are randomly chosen from a uniform distribution over the range of the data. Optionally, a random number seed may be specified to cause the command set seed *seed#* (see [R] **generate**) to be applied before the k group centers are generated.

generate(*groupvar*) provides the name of the grouping variable to be created by cluster kmeans. By default, it will be the name specified in name().

iterate(*#*) specifies the maximum number of iterations to allow in the kmeans clustering algorithm. The default is iterate(10000).

keepcenters indicates that the group means, from the k groups that are produced, are to be appended to the data.

Remarks

Two examples of using cluster kmeans are presented, one using continuous data and the other using binary data. See [CL] **cluster kmedians** to see these same two datasets examined using kmedians clustering.

▷ Example

You have measured the flexibility, speed, and strength of the 80 students in your physical education class. You want to split the class into four groups, based on their physical attributes, so that they can receive the mix of flexibility, strength, and speed training that will best help them improve.

Here is a summary of the data and a matrix graph showing the data:

```
. use http://www.stata-press.com/data/r8/physed
. summarize flex speed strength
```

Variable	Obs	Mean	Std. Dev.	Min	Max
flexibility	80	4.402625	2.788541	.03	9.97
speed	80	3.875875	3.121665	.03	9.79
strength	80	6.439875	2.449293	.05	9.57

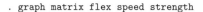

. graph matrix flex speed strength

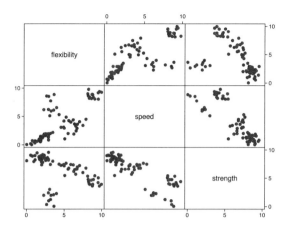

As you expected, based on what you saw the first day of class, the data indicate a wide range of levels of performance for the students. The graph seems to indicate that there are some distinct groups, which leads you to believe that your plan will work well.

You decide to perform a cluster analysis to create four groups, one for each of your class assistants. You have had good experience with kmeans clustering in the past, and generally like the behavior of the absolute-value distance.

You don't really care what starting values are used in the cluster analysis, but you do want to be able to reproduce the same results if you ever decide to rerun your analysis. You decide to use the krandom() option to pick k of the observations at random as the initial group centers. You supply a random number seed for reproducibility. You also add the keepcenters option so that the means of the four groups will be added to the bottom of your dataset.

```
. cluster k flex speed strength, k(4) name(g4abs) start(kr(385617)) abs keepcen
. cluster list g4abs
g4abs  (type: partition,  method: kmeans,  dissimilarity: L1)
     vars: g4abs (group variable)
    other: k: 4
          start: krandom(385617)
          range: 0 .
          cmd: cluster kmeans flex speed strength, k(4) name(g4abs)
               start(kr(385617)) abs keepcen
          varlist: flexibility speed strength

. table g4abs
```

g4abs	Freq.
1	15
2	20
3	35
4	10

```
. list flex speed strength in 81/1
```

	flexib~y	speed	strength
81.	8.852	8.743333	4.358
82.	5.9465	3.4485	6.8325
83.	1.969429	1.144857	8.478857
84.	3.157	6.988	1.641

```
. drop in 81/1
(4 observations deleted)
. tabstat flex speed strength, by(g4abs) stat(min mean max)
```

Summary statistics: min, mean, max
  by categories of: g4abs

g4abs	flexib~y	speed	strength
1	8.12	8.05	3.61
	8.852	8.743333	4.358
	9.97	9.79	5.42
2	4.32	1.05	5.46
	5.9465	3.4485	6.8325
	7.89	5.32	7.66
3	.03	.03	7.38
	1.969429	1.144857	8.478857
	3.48	2.17	9.57
4	2.29	5.11	.05
	3.157	6.988	1.641
	3.99	8.87	3.02
Total	.03	.03	.05
	4.402625	3.875875	6.439875
	9.97	9.79	9.57

After looking at the last four observations (which are the group means since you specified keepcenters), you decided that what you really wanted to see was the minimum and maximum values and the mean for the four groups. You removed the last four observations, and then used the tabstat command to view the desired statistics.

Group 1, with 15 students, is already doing very well in flexibility and speed, but will need extra strength training. Group 2, with 20 students, needs to emphasize speed training, but could use some improvement in the other categories as well. Group 3, the largest, with 35 students, has serious problems with both flexibility and speed, though they did very well in the strength category. Group 4, the smallest, with 10 students, needs help with flexibility and strength.

Since you like looking at graphs, you decide to view the matrix graph again, but with the addition of group numbers as plotting symbols.

(Continued on next page)

. graph matrix flex speed strength, m(i) mlabel(g4abs) mlabpos(0)

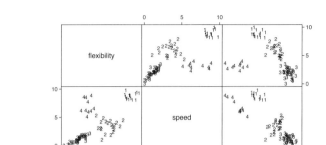

The groups, as shown in the graph, do appear reasonably distinct. However, you had hoped to have groups that were about the same size. You are curious what clustering to three or five groups would produce. For no good reason, you decide to use the first k observations as initial group centers for clustering to three groups, and random numbers within the range of the data for clustering to five groups.

. cluster k flex speed strength, k(3) name(g3abs) start(firstk) abs
. cluster k flex speed strength, k(5) name(g5abs) start(random(33576)) abs
. table g3abs g4abs, col

			g4abs		
g3abs	1	2	3	4	Total
1				10	10
2		18	35		53
3	15	2			17

. table g5abs g4abs, col

			g4abs		
g5abs	1	2	3	4	Total
1		20			20
2	15				15
3				6	6
4				4	4
5			35		35

With three groups, the unequal-group-size problem gets worse. With five groups, the smallest group gets split. Four groups seem like the best option for this class. You will try to help the assistant assigned to group 3 in dealing with the larger group.

◁

▷ Example

You have just started a women's club. Thirty women, from throughout the community, have sent in their requests to join. You have them fill out a questionnaire with 35 yes/no questions relating to sports, music, reading, and hobbies. Here is a description of the dataset:

```
. use http://www.stata-press.com/data/r8/wclub

. describe
```

Contains data from http://www.stata-press.com/data/r8/wclub.dta

obs:	30		
vars:	35		8 Nov 2002 14:06
size:	1,170 (99.5% of memory free)		

variable name	storage type	display format	value label	variable label
bike	byte	%8.0g		enjoy bicycle riding Y/N
bowl	byte	%8.0g		enjoy bowling Y/N
swim	byte	%8.0g		enjoy swimming Y/N
jog	byte	%8.0g		enjoy jogging Y/N
hock	byte	%8.0g		enjoy watching hockey Y/N
foot	byte	%8.0g		enjoy watching football Y/N
base	byte	%8.0g		enjoy baseball Y/N
bask	byte	%8.0g		enjoy basketball Y/N
arob	byte	%8.0g		participate in aerobics Y/N
fshg	byte	%8.0g		enjoy fishing Y/N
dart	byte	%8.0g		enjoy playing darts Y/N
clas	byte	%8.0g		enjoy classical music Y/N
cntr	byte	%8.0g		enjoy country music Y/N
jazz	byte	%8.0g		enjoy jazz music Y/N
rock	byte	%8.0g		enjoy rock and roll music Y/N
west	byte	%8.0g		enjoy reading western novels Y/N
romc	byte	%8.0g		enjoy reading romance novels Y/N
scif	byte	%8.0g		enjoy reading sci. fiction Y/N
biog	byte	%8.0g		enjoy reading biographies Y/N
fict	byte	%8.0g		enjoy reading fiction Y/N
hist	byte	%8.0g		enjoy reading history Y/N
cook	byte	%8.0g		enjoy cooking Y/N
shop	byte	%8.0g		enjoy shopping Y/N
soap	byte	%8.0g		enjoy watching soap operas Y/N
sew	byte	%8.0g		enjoy sewing Y/N
crft	byte	%8.0g		enjoy craft activities Y/N
auto	byte	%8.0g		enjoy automobile mechanics Y/N
pokr	byte	%8.0g		enjoy playing poker Y/N
brdg	byte	%8.0g		enjoy playing bridge Y/N
kids	byte	%8.0g		have children Y/N
hors	byte	%8.0g		have a horse Y/N
cat	byte	%8.0g		have a cat Y/N
dog	byte	%8.0g		have a dog Y/N
bird	byte	%8.0g		have a bird Y/N
fish	byte	%8.0g		have a fish Y/N

Sorted by:

Now you are trying to plan the first club meeting. You decide to have a lunch along with the business meeting that will officially organize the club and ratify its charter. You want the club to get off to a good start, so you worry about the best way to seat the guests. You decide to use kmeans clustering on the yes/no data from the questionnaires to put people with similar interests at the same tables.

You have five tables that can each seat up to eight comfortably. You request clustering to five groups and hope that the group sizes will fall under this table size limit.

You really want people placed together based on shared positive interests, instead of on shared noninterests. From among all the available binary similarity measures, you decide to use the Jaccard coefficient, since it does not include jointly zero comparisons in its formula; see [CL] **cluster**. The Jaccard coefficient is also easy to understand.

```
. cluster kmeans bike-fish, k(5) Jaccard st(firstk) name(gr5)
. cluster list gr5
gr5 (type: partition,  method: kmeans,  similarity: Jaccard)
      vars: gr5 (group variable)
     other: k: 5
            start: firstk
            range: 1 0
            cmd: cluster kmeans bike-fish, k(5) Jaccard st(firstk) name(gr5)
            varlist: bike bowl swim jog hock foot base bask arob fshg dart clas
                cntr jazz rock west romc scif biog fict hist cook shop soap
                sew crft auto pokr brdg kids hors cat dog bird fish
. table gr5
```

gr5	Freq.
1	7
2	7
3	5
4	5
5	6

You get lucky; the groups are reasonably close in size. You will seat yourself at one of the tables with only five people, and your sister, who did not fill out a questionnaire, at the other table with only five people to make things as even as possible.

Now, you wonder, what are the characteristics of these five groups? You decide to use the `tabstat` command to view the proportion answering yes to each question for each of the five groups.

```
. tabstat bike-fish, by(gr5) format(%4.3f)
Summary statistics: mean
  by categories of: gr5
```

gr5	bike	bowl	swim	jog	hock	foot
1	0.714	0.571	0.714	0.571	0.143	0.143
2	0.286	0.143	0.571	0.714	0.143	0.143
3	0.400	0.200	0.600	0.200	0.200	0.400
4	0.200	0.000	0.200	0.200	0.000	0.400
5	0.000	0.500	0.000	0.000	0.333	0.167
Total	0.333	0.300	0.433	0.367	0.167	0.233

gr5	base	bask	arob	fshg	dart	clas
1	0.429	0.571	0.857	0.429	0.571	0.429
2	0.571	0.286	0.714	0.429	0.857	0.857
3	0.600	0.400	0.000	0.800	0.200	0.000
4	0.200	0.600	0.400	0.000	0.000	0.800
5	0.167	0.333	0.000	0.500	0.167	0.000
Total	0.400	0.433	0.433	0.433	0.400	0.433

gr5	cntr	jazz	rock	west	romc	scif
1	0.857	0.571	0.286	0.714	0.571	0.286
2	0.571	0.857	0.429	0.143	0.143	0.857
3	0.200	0.200	0.600	0.000	0.000	0.200
4	0.200	0.400	0.400	0.200	0.400	0.000
5	0.833	0.167	0.667	0.500	0.667	0.000
Total	0.567	0.467	0.467	0.333	0.367	0.300

gr5	biog	fict	hist	cook	shop	soap
1	0.429	0.429	0.571	0.714	0.571	0.571
2	0.429	0.571	0.571	0.000	0.429	0.143
3	0.000	0.200	0.000	0.600	1.000	0.600
4	1.000	1.000	1.000	0.600	0.600	0.200
5	0.000	0.167	0.000	0.333	1.000	0.667
Total	0.367	0.467	0.433	0.433	0.700	0.433

gr5	sew	crft	auto	pokr	brdg	kids
1	0.429	0.571	0.143	0.571	0.429	0.714
2	0.143	0.714	0.429	0.286	0.714	0.143
3	0.400	0.200	0.600	1.000	0.200	0.600
4	0.800	0.800	0.000	0.000	0.000	1.000
5	0.000	0.000	0.333	0.667	0.000	0.500
Total	0.333	0.467	0.300	0.500	0.300	0.567

gr5	hors	cat	dog	bird	fish	
1	0.571	0.571	1.000	0.286	0.429	
2	0.143	0.571	0.143	0.429	0.143	
3	0.000	0.200	0.200	0.400	0.800	
4	0.000	0.400	0.000	0.000	0.200	
5	0.167	0.167	0.833	0.167	0.167	
Total	0.200	0.400	0.467	0.267	0.333	

It appears that group 1 likes participating in most sporting activities, prefers country music, likes reading western and romance novels, enjoys cooking, and is more likely to have kids and various animals, including horses.

Group 2 likes some sports (swimming, jogging, aerobics, baseball, and darts), prefers classical and jazz music, prefers science fiction (but also enjoys biography, fiction, and history), dislikes cooking, enjoys playing bridge, is not likely to have children, and is more likely to have a cat than any other animal.

Group 3 seems to enjoy swimming, baseball, and fishing (but dislikes aerobics), prefers rock and roll music (disliking classical), does not enjoy reading, prefers poker over bridge, and is more likely to own a fish than any other animal.

Group 4 dislikes many of the sports, prefers classical music, likes reading biographies, fiction, and history, enjoys sewing and crafts, dislikes card games, has kids, and is not likely to have pets.

Group 5 dislikes sports, prefers country and rock and roll music, will pick up romance and western novels on occasion, dislikes sewing and crafts, prefers poker instead of bridge, and is most likely to have a dog.

◁

Methods and Formulas

Kmeans cluster analysis is discussed in most cluster analysis books; see the references in [CL] **cluster**. [CL] **cluster** also provides a general discussion of cluster analysis, including kmeans clustering, and discusses the available `cluster` subcommands.

Kmeans clustering is an iterative procedure that partitions the data into k groups or clusters. The procedure begins with k initial group centers. Observations are assigned to the group with the closest center. The mean of the observations assigned to each of the groups is computed, and the process is repeated. These steps continue until all observations remain in the same group from the previous iteration.

To avoid endless loops, an observation will only be reassigned to a different group if it is closer to the other group center. In the case of a tied distance between an observation and two or more group centers, the observation is assigned to its current group if that is one of the closest, and to the lowest numbered group otherwise.

The `start()` option provides many ways of specifying the beginning group centers. These include methods that specify the actual starting centers, as well as methods that specify initial partitions of the data from which the beginning centers are computed.

Some kmeans clustering algorithms recompute the group centers after each reassignment of an observation. Other kmeans clustering algorithms, including Stata's `cluster kmeans` command, recompute the group centers only after a complete pass through the data. A disadvantage of this method is that orphaned group centers can occur. An orphaned center is one that has no observations that are closest to it. The advantage of recomputing means only at the end of each pass through the data is that the sort order of the data does not potentially change your final result.

Stata deals with orphaned centers by finding the observation that is farthest from its center and using that as a new group center. The observations are then reassigned to the closest groups, including this (these) new center(s).

Continuous or binary data are allowed with `cluster kmeans`. The mean of a group of binary observations for a variable is the proportion of ones for that group of observations and variable. The binary similarity measures can accommodate the comparison of a binary observation to a binary mean (proportion). See [CL] **cluster** for details on this subject and for the formulas for all the available (dis)similarity measures.

Also See

Complementary:	[CL] **cluster notes**, [CL] **cluster stop**, [CL] **cluster utility**
Related:	[CL] **cluster kmedians**
Background:	[CL] **cluster**

Title

cluster kmedians — Kmedians cluster analysis

Syntax

cluster <u>kmed</u>ians [*varlist*] [if *exp*] [in *range*] , k(#) [<u>name</u>(*clname*)

    distance_option <u>s</u>tart(*start_option*) <u>gen</u>erate(*groupvar*) <u>iter</u>ate(#) <u>keep</u>centers]

Description

cluster kmedians performs kmedians partition cluster analysis. See [CL] **cluster** for a general discussion of cluster analysis and for a description of the other cluster commands. See [CL] **cluster kmeans** for an alternative that uses means instead of medians.

Options

k(#) is required, and indicates that # groups are to be formed by the cluster analysis.

name(*clname*) specifies the name to attach to the resulting cluster analysis. If name() is not specified, Stata finds an available cluster name, displays it for your reference, and then attaches the name to your cluster analysis.

distance_option is one of the similarity or dissimilarity measures allowed by Stata. Capitalization of the option does not matter. See [CL] **cluster** for a discussion of these measures.

    The available measures designed for continuous data are L2 (synonym <u>Euc</u>lidean), which is the default; L2squared; L1 (synonyms <u>absolute</u>, cityblock, and <u>manhattan</u>); <u>L</u>infinity (synonym <u>maximum</u>); L(#); <u>Lpow</u>er(#); <u>Can</u>berra; <u>corr</u>elation; and <u>ang</u>ular (synonym angle).

    The available measures designed for binary data are <u>match</u>ing, Jaccard, <u>Russ</u>ell, Hamman, Dice, antiDice, Sneath, Rogers, Ochiai, Yule, <u>Ander</u>berg, <u>Kulc</u>zynski, Gower2, and Pearson.

start(*start_option*) indicates how the k initial group centers are to be obtained. The available *start_option*s are <u>kr</u>andom[(*seed#*)], <u>firstk</u>[, <u>ex</u>clude], <u>lastk</u>[, <u>ex</u>clude], <u>prandom</u>[(*seed#*)], <u>everyk</u>th, <u>segments</u>, <u>group</u>(*varname*), and <u>r</u>andom[(*seed#*)].

    krandom[(*seed#*)], the default, indicates that k unique observations are to be chosen at random, from among those to be clustered, as starting centers for the k groups. Optionally, a random number seed may be specified to cause the command set seed *seed#* (see [R] **generate**) to be applied before the k random observations are chosen.

    firstk[, exclude] indicates that the first k observations, from among those to be clustered, are to be used as the starting centers for the k groups. With the addition of the exclude option, these first k observations are then not included among the observations to be clustered.

    lastk[, exclude] indicates that the last k observations, from among those to be clustered, are to be used as the starting centers for the k groups. With the addition of the exclude option, these last k observations are then not included among the observations to be clustered.

prandom$\big[(seed\#)\big]$ indicates that k partitions are to be formed randomly among the observations to be clustered. The group medians from the k groups defined by this partitioning are used as the starting group centers. Optionally, a random number seed may be specified to cause the command set seed *seed#* (see [R] **generate**) to be applied before the k partitions are chosen.

everykth indicates that k partitions are to be formed by assigning observations 1, $1+k$, $1+2k$, ... to the first group; assigning observations 2, $2+k$, $2+2k$, ... to the second group; and so on, to form k groups. The group medians from these k groups are used as the starting group centers.

segments indicates that k nearly equal partitions are to be formed from the data. Approximately the first N/k observations are assigned to the first group, the second N/k observations are assigned to the second group, and so on. The group medians from these k groups are used as the starting group centers.

group(*varname*) provides an initial grouping variable, *varname*, that defines k groups among the observations to be clustered. The group medians from these k groups are used as the starting group centers.

random$\big[(seed\#)\big]$ indicates that k random initial group centers are to be generated. The values are randomly chosen from a uniform distribution over the range of the data. Optionally, a random number seed may be specified to cause the command set seed *seed#* (see [R] **generate**) to be applied before the k group centers are generated.

generate(*groupvar*) provides the name of the grouping variable to be created by cluster kmedians. By default, it will be the name specified in name().

iterate(#) specifies the maximum number of iterations to allow in the kmedians clustering algorithm. The default is iterate(10000).

keepcenters indicates that the group medians, from the k groups that are produced, are to be appended to the data.

Remarks

The data from the two examples introduced in [CL] **cluster kmeans** are presented here to demonstrate the use of cluster kmedians. The first dataset contains continuous data, and the second dataset contains binary data.

▷ Example

You have measured the flexibility, speed, and strength of the 80 students in your physical education class. You want to split the class into four groups, based on their physical attributes, so that they can receive the mix of flexibility, strength, and speed training that will best help them improve.

The data are summarized and graphed in [CL] **cluster kmeans**. You previously performed kmeans clustering on these data to obtain three, four, and five groups.

```
. use http://www.stata-press.com/data/r8/physed
. cluster k flex speed strength, k(4) name(g4abs) abs start(kr(385617))
. cluster k flex speed strength, k(3) name(g3abs) abs start(firstk)
. cluster k flex speed strength, k(5) name(g5abs) abs start(random(33576))
```

You now wish to see if kmedians clustering will produce the same grouping for this dataset. You specify four groups, absolute-value distance, and k random observations as beginning centers (but using a different random number seed).

```
. cluster kmed flex speed strength, k(4) name(kmed4) abs start(kr(11736))
```

```
. cluster list kmed4
kmed4 (type: partition, method: kmedians, dissimilarity: L1)
     vars: kmed4 (group variable)
    other: k: 4
           start: krandom(11736)
           range: 0 .
           cmd: cluster kmedians flex speed strength, k(4) name(kmed4) abs
                start(kr(11736))
           varlist: flexibility speed strength

. table g4abs kmed4
```

g4abs	kmed4 1	2	3	4
1		15		
2			20	
3	35			
4				10

Other than a difference in how the groups are numbered, kmedians clustering and kmeans clustering produced the same results for this dataset.

Now you want to see what happens with kmedians clustering for three groups and for five groups.

```
. cluster kmed flex speed strength, k(3) name(kmed3) abs start(lastk)
. cluster kmed flex speed strength, k(5) name(kmed5) abs start(prand(8723))
. cluster list kmed3 kmed5
kmed3 (type: partition, method: kmedians, dissimilarity: L1)
     vars: kmed3 (group variable)
    other: k: 3
           start: lastk
           range: 0 .
           cmd: cluster kmedians flex speed strength, k(3) name(kmed3) abs
                start(lastk)
           varlist: flexibility speed strength
kmed5 (type: partition, method: kmedians, dissimilarity: L1)
     vars: kmed5 (group variable)
    other: k: 5
           start: prandom(8723)
           range: 0 .
           cmd: cluster kmedians flex speed strength, k(5) name(kmed5) abs
                start(prand(8723))
           varlist: flexibility speed strength

. table g3abs kmed3, row
```

g3abs	kmed3 1	2	3
1	6		4
2	18	35	
3	2		15
Total	26	35	19

. table g5abs kmed5, row

			kmed5		
g5abs	1	2	3	4	5
1		20			
2	15				
3				6	
4				4	
5			20		15
Total	15	20	20	10	15

Kmeans and kmedians clustering produced different groups for three groups and five groups.

Since one of your concerns was having a better balance in the group sizes, you decide to look a little bit closer at the five-group solution produced by kmedians clustering.

. table g4abs kmed5, row col

			kmed5			
g4abs	1	2	3	4	5	Total
1	15					15
2		20				20
3			20		15	35
4				10		10
Total	15	20	20	10	15	80

. tabstat flex speed strength, by(kmed5) stat(min mean max)

Summary statistics: min, mean, max
  by categories of: kmed5

kmed5	flexib~y	speed	strength
1	8.12	8.05	3.61
	8.852	8.743333	4.358
	9.97	9.79	5.42
2	4.32	1.05	5.46
	5.9465	3.4485	6.8325
	7.89	5.32	7.66
3	1.85	1.18	7.38
	2.4425	1.569	8.2775
	3.48	2.17	9.19
4	2.29	5.11	.05
	3.157	6.988	1.641
	3.99	8.87	3.02
5	.03	.03	7.96
	1.338667	.5793333	8.747333
	2.92	.99	9.57
Total	.03	.03	.05
	4.402625	3.875875	6.439875
	9.97	9.79	9.57

A matrix graph with group numbers as plotting symbols helps you visualize the data.

. graph matrix flex speed strength, m(i) mlabel(kmed5) mlabpos(0)

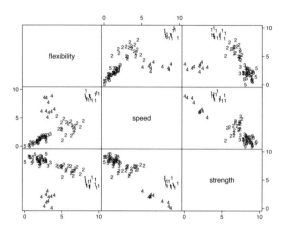

The five-group kmedians clustering split the group with 35 students from the four-group clustering into groups of size 20 and size 15. Looking at the output from tabstat, you see that this group was broken up so that the 15 slowest students are split apart from the other 20 (who are still slower than the remaining groups).

The characteristics of the five groups are as follows: Group 1, with 15 students, is already doing very well in flexibility and speed, but will need extra strength training. Group 2, with 20 students, needs to emphasize speed training, but could use some improvement in the other categories as well. Group 3, which used to have 35 students, now has 20 students, and has serious problems with both flexibility and speed, though they did very well in the strength category. Group 4, with 10 students, needs help with flexibility and strength. Group 5, which was split off of Group 3, has the 15 slowest students.

Even though the matrix graph showing the five groups does not indicate that groups 3 and 5 are very distinct, you decide to go with five groups anyway to even out the group sizes. You will take the slowest group and work with them directly, since they will need a lot of extra help, while your four class assistants will take care of the other four groups.

◁

▷ Example

As explained in the second example of [CL] **cluster kmeans**, you have just started a women's club. Thirty women, from throughout the community, have sent in their requests to join. You have them fill out a questionnaire with 35 yes/no questions relating to sports, music, reading, and hobbies. A description of the data is found in [CL] **cluster kmeans**.

In planning the first meeting of the club, you are trying to assign seats at the five lunch tables based on shared interests among the women. Kmeans clustering gave you five groups that are each about the right size and seem to make sense in terms of each groups common characteristics; see [CL] **cluster kmeans**. Now you want to see if kmedian clustering suggests a better solution.

As before, you select the Jaccard coefficient as the binary similarity measure, and the first k observations as starting centers.

```
. use http://www.stata-press.com/data/r8/wclub
. cluster kmeans bike-fish, k(5) Jaccard st(firstk) name(gr5)
. cluster kmed bike-fish, k(5) Jaccard st(firstk) name(kmedian5)
. cluster list kmedian5
kmedian5 (type: partition,  method: kmedians,  similarity: Jaccard)
     vars: kmedian5 (group variable)
    other: k: 5
           start: firstk
           range: 1 0
           cmd: cluster kmedians bike-fish, k(5) Jaccard st(firstk)
               name(kmedian5)
           varlist: bike bowl swim jog hock foot base bask arob fshg dart clas
                   cntr jazz rock west romc scif biog fict hist cook shop soap
                   sew crft auto pokr brdg kids hors cat dog bird fish
```

```
. table gr5 kmedian5, row col
```

			kmedian5			
gr5	1	2	3	4	5	Total
1	7					7
2	1	6				7
3			5			5
4				5		5
5	1		1		4	6
Total	9	6	6	5	4	30

The resulting groups are similar. Three ladies are grouped differently with this kmedian clustering compared with the kmeans clustering. However, there is a more even distribution of women to the five groups of the kmeans clustering. Since the five lunch tables can seat only eight comfortably, and the kmedians clustering produces one group of size nine, you decide to stick with the groups produced by kmeans clustering.

◁

Methods and Formulas

Kmedians cluster analysis is a variation of the standard kmeans clustering discussed in most cluster analysis books; see the references in [CL] **cluster**. [CL] **cluster** also provides a general discussion of cluster analysis, including kmeans and kmedians clustering, and discusses the available cluster subcommands.

Kmedians clustering is an iterative procedure that partitions the data into k groups or clusters. The procedure begins with k initial group centers. Observations are assigned to the group with the closest center. The median of the observations assigned to each of the groups is computed, and the process is repeated. These steps continue until all observations remain in the same group from the previous iteration.

To avoid endless loops, an observation will only be reassigned to a different group if it is closer to the other group center. In the case of a tied distance between an observation and two or more group centers, the observation is assigned to its current group if that is one of the closest, or to the lowest numbered group otherwise.

The start() option provides many ways of specifying the beginning group centers. These include methods that specify the actual starting centers, as well as methods that specify initial partitions of the data from which the beginning centers are computed.

Stata's `cluster kmedians` command recomputes the group centers only after a complete pass through the data. A disadvantage of this method is that orphaned group centers can occur. An orphaned center is one that has no observations that are closest to it. The advantage of recomputing means only at the end of each pass through the data, instead of after each reassignment, is that the sort order of the data does not potentially change your final result.

Stata deals with orphaned centers by finding the observation that is farthest from its center and using that as a new group center. The observations are then reassigned to the closest groups, including this (these) new center(s).

Continuous or binary data are allowed with `cluster kmedians`. The median of a group of binary observations for a variable is almost always either zero or one. However, if there are an equal number of zeros and ones for a group, then the median is 0.5, which is treated as a proportion (just as with kmeans clustering). The binary similarity measures can accommodate the comparison of a binary observation to a proportion. See [CL] **cluster** for details on this subject and for the formulas for all the available (dis)similarity measures.

Also See

Complementary:	[CL] **cluster notes**, [CL] **cluster stop**, [CL] **cluster utility**
Related:	[CL] **cluster kmeans**
Background:	[CL] **cluster**

Title

cluster medianlinkage — Median linkage cluster analysis

Syntax

cluster <u>median</u>linkage [*varlist*] [if *exp*] [in *range*] [, <u>na</u>me(*clname*)

    distance_option <u>gen</u>erate(*stub*)]

Description

The cluster medianlinkage command performs hierarchical agglomerative median linkage cluster analysis. See [CL] **cluster** for a general discussion of cluster analysis and for a description of the other cluster commands. The cluster dendrogram command (see [CL] **cluster dendrogram**) will display the resulting dendrogram, the cluster stop command (see [CL] **cluster stop**) will help in determining the number of groups, and the cluster generate command (see [CL] **cluster generate**) will produce grouping variables.

Options

<u>na</u>me(*clname*) specifies the name to attach to the resulting cluster analysis. If name() is not specified, Stata finds an available cluster name, displays it for your reference, and then attaches the name to your cluster analysis.

distance_option is one of the similarity or dissimilarity measures allowed by Stata. Capitalization of the option does not matter. See [CL] **cluster** for a discussion of these measures.

The available measures designed for continuous data are L2 (synonym <u>Euc</u>lidean); L2squared, which is the default for cluster medianlinkage; L1 (synonyms <u>abs</u>olute, <u>city</u>block, and <u>manhat</u>tan); <u>Linf</u>inity (synonym <u>max</u>imum); L(*#*); <u>Lpow</u>er(*#*); <u>Can</u>berra; <u>corr</u>elation; and <u>ang</u>ular (synonym <u>angle</u>).

The available measures designed for binary data are <u>mat</u>ching, Jaccard, <u>Russ</u>ell, Hamman, Dice, antiDice, Sneath, Rogers, Ochiai, Yule, <u>Ander</u>berg, <u>Kulc</u>zynski, Gower2, and Pearson.

Several authors advise the exclusive use of the L2squared *distance_option* with median linkage. See the sections *(Dis)similarity transformations* and *the Lance and Williams formula* and *Warning concerning (dis)similarity choice* in [CL] **cluster** for details.

<u>gen</u>erate(*stub*) provides a prefix for the variable names created by cluster medianlinkage. By default, the variable-name prefix will be the name specified in name(). Three variables are created and attached to the cluster analysis results, with the suffixes _id, _ord, and _hgt. Users generally will not need to access these variables directly.

Median linkage can produce reversals or crossovers; see [CL] **cluster** for details. When reversals happen, cluster medianlinkage also creates a fourth variable with the suffix _pht. This is a pseudo-height variable that is used by some of the post-clustering commands to properly interpret the _hgt variable.

Remarks

An example using the default L2squared (squared Euclidean) distance and L2 (Euclidean) distance on continuous data and an example using the matching coefficient on binary data illustrate the cluster medianlinkage command. These are the same datasets introduced in [CL] **cluster singlelinkage**, which are used as examples for all the hierarchical clustering methods, so that you can compare the results from using different hierarchical clustering methods.

▷ Example

As explained in the first example of [CL] **cluster singlelinkage**, as the senior data analyst for a small biotechnology firm, you are given a dataset with 4 chemical laboratory measurements on 50 different samples of a particular plant gathered from the rain forest. The head of the expedition that gathered the samples thinks, based on information from the natives, that an extract from the plant might reduce the negative side effects associated with your company's best-selling nutritional supplement.

While the company chemists and botanists continue exploring the possible uses of the plant and plan future experiments, the head of product development asks you to look at the preliminary data and to report anything that might be helpful to the researchers.

While all 50 of the plants are supposed to be of the same type, you decide to perform a cluster analysis to see if there are subgroups or anomalies among them. Single linkage clustering helped you discover an anomaly in the data. You now wish to see if you discover the same thing using median linkage clustering with the default squared Euclidean distance and with Euclidean distance.

You first call cluster medianlinkage, letting the distance default to L2squared (squared Euclidean distance), and use the name() option to attach the name med to the resulting cluster analysis. The cluster list command (see [CL] **cluster utility**) is then applied to list the components of your cluster analysis.

```
. use http://www.stata-press.com/data/r8/labtech
. cluster medianlinkage x1 x2 x3 x4, name(med)
. cluster list med
med  (type: hierarchical,  method: median,  dissimilarity: L2squared)
       vars: med_id (id variable)
             med_ord (order variable)
             med_hgt (real_height variable)
             med_pht (pseudo_height variable)
      other: range: 0 .
             cmd: cluster medianlinkage x1 x2 x3 x4, name(med)
             varlist: x1 x2 x3 x4
```

You do the same thing again, but this time using L2 (Euclidean distance) and giving it the name L2med.

```
. cluster medianlinkage x1 x2 x3 x4, name(L2med) L2
. cluster list L2med
L2med  (type: hierarchical,  method: median,  dissimilarity: L2)
        vars: L2med_id (id variable)
              L2med_ord (order variable)
              L2med_hgt (real_height variable)
              L2med_pht (pseudo_height variable)
       other: range: 0 .
              cmd: cluster medianlinkage x1 x2 x3 x4, name(L2med) L2
              varlist: x1 x2 x3 x4
```

You wish to use the `cluster dendrogram` command to graph the dendrogram (see [CL] **cluster dendrogram**), but since this particular cluster analysis produces reversals, it refuses to draw the dendrogram.

You decide to use the `cluster generate` command (see [CL] **cluster generate**) to produce grouping variables for 2 to 10 groups for each of the two cluster analyses. You also wish to examine the cross-tabulation of each of these generated groups against a variable that identifies which laboratory technician produced the data. (For the sake of brevity, only one cross-tabulation is shown for each of the cluster analyses.)

```
. cluster gen gm = groups(2/10), name(med)
. cluster gen gL2m = groups(2/10), name(L2med)
. table labtech gm2
```

labtech	gm2 1	2
Al	10	
Bill	10	
Deb	10	
Jen	10	
Sam		10

```
. table labtech gL2m4
```

labtech	gL2m4 1	2	3	4
Al	10			
Bill	10			
Deb	8		1	1
Jen	10			
Sam		10		

The samples analyzed by Sam appear to want to stay clustered together more strongly than the samples analyzed by the other technicians. The reason for this phenomenon is not as obvious from this analysis as it was when viewing the dendrogram from the single linkage clustering (see [CL] **cluster singlelinkage**).

◁

▷ Example

This example analyzes the same data as introduced in the second example of [CL] **cluster singlelinkage**. The sociology professor of your graduate-level class gives, as homework, a dataset containing 30 observations on 60 binary variables, with the assignment to tell him something about the 30 subjects represented by the observations.

In addition to examining single linkage clustering of these data, you decide to see what median linkage clustering shows. As with the single linkage clustering, you pick the simple matching binary coefficient to measure the similarity between groups. The `name()` option is used to attach the name `medlink` to the cluster analysis. `cluster list` displays the details; see [CL] **cluster utility**.

```
. use http://www.stata-press.com/data/r8/homework
. cluster median a1-a60, match name(medlink)
. cluster list medlink
medlink (type: hierarchical,  method: median,   similarity: matching)
      vars: medlink_id (id variable)
            medlink_ord (order variable)
            medlink_hgt (real_height variable)
            medlink_pht (pseudo_height variable)
     other: range: 1 0
            cmd: cluster medianlinkage a1-a60, match name(medlink)
            varlist: a1 a2 a3 a4 a5 a6 a7 a8 a9 a10 a11 a12 a13 a14 a15 a16 a17
                  a18 a19 a20 a21 a22 a23 a24 a25 a26 a27 a28 a29 a30 a31 a32
                  a33 a34 a35 a36 a37 a38 a39 a40 a41 a42 a43 a44 a45 a46 a47
                  a48 a49 a50 a51 a52 a53 a54 a55 a56 a57 a58 a59 a60
```

You attempt to use the `cluster dendrogram` command to display the dendrogram, but since this particular cluster analysis produced reversals, `cluster dendrogram` refuses to produce the dendrogram. You realize that with reversals the resulting dendrogram would not be easy to interpret anyway.

You decide to go directly to comparing the three-group solution from this median linkage clustering with the variable called `truegrp` provided by the teacher. You use the `cluster generate` command (see [CL] **cluster generate**) to create a grouping variable, based on your centroid clustering, to compare with `truegrp`.

```
. cluster gen medgrp3 = group(3)
. table medgrp3 truegrp
```

medgrp3	truegrp 1	2	3
1		10	
2	10		
3			10

Other than the numbers arbitrarily assigned to the three groups, your teacher's conclusions and the results from the three-group median linkage clustering are in complete agreement.

◁

❑ Technical Note

`cluster medianlinkage` requires more memory and more execution time than `cluster singlelinkage`. With a large number of observations, the execution time may be significant.

❑

Methods and Formulas

[CL] **cluster** discusses hierarchical clustering, and places median linkage clustering in this general framework. Conceptually, hierarchical agglomerative clustering proceeds as follows. The N observations start out as N separate groups, each of size one. The two closest observations are merged into one group, producing $N-1$ total groups. The closest two groups are then merged, so that there are $N-2$ total groups. This process continues until all the observations are merged into one large group. This produces a hierarchy of groupings from one group to N groups. The difference between the various hierarchical linkage methods depends on how "closest" is defined when comparing groups.

Median linkage clustering is a variation on centroid linkage clustering. The difference is in how groups of unequal size are treated. Centroid linkage gives each observation equal weight. Median linkage gives each group of observations equal weight, meaning that with unequal group sizes, the observations in the smaller group will have more weight than the observations in the larger group.

The median linkage clustering algorithm produces two variables that act as a pointer representation of a dendrogram. To this, Stata adds a third variable used to restore the sort order, as needed, so that the two variables of the pointer representation remain valid. The first variable of the pointer representation gives the order of the observations. The second variable has one less element, and gives the height in the dendrogram at which the adjacent observations in the order-variable join. When reversals happen, which they often do, a fourth variable, called a pseudo-height, is produced. This is used by post-clustering commands in conjunction with the height variable to properly interpret the ordering of the hierarchy.

See [CL] **cluster** for the details, warnings, and formulas of the available *distance_option*s, which include (dis)similarity measures for continuous and for binary data.

Also See

Complementary:	[CL] **cluster dendrogram**, [CL] **cluster generate**, [CL] **cluster notes**, [CL] **cluster stop**, [CL] **cluster utility**
Related:	[CL] **cluster averagelinkage**, [CL] **cluster centroidlinkage**, [CL] **cluster completelinkage**, [CL] **cluster singlelinkage**, [CL] **cluster wardslinkage**, [CL] **cluster waveragelinkage**
Background:	[CL] **cluster**

Title

cluster notes — Place notes in cluster analysis

Syntax

cluster <u>notes</u> *clname* : *text*

cluster <u>notes</u>

cluster <u>notes</u> *clnamelist*

cluster <u>notes</u> drop *clname* [in *numlist*]

Description

The cluster notes command attaches notes to a previously run cluster analysis. The notes become part of the data, and are saved when the data are saved and retrieved when the data are used; see [R] **save**.

To add a note to a cluster analysis, type cluster notes, the cluster analysis name, a colon, and the text.

Typing cluster notes by itself will list all cluster notes associated with all defined cluster analyses. cluster notes followed by one or more cluster names lists the notes for those cluster analyses.

cluster notes drop allows you to drop cluster notes.

Remarks

The cluster analysis system in Stata has many features that allow you to manage the various cluster analyses that you perform. See [CL] **cluster** for information on all the available cluster analysis commands, and, in particular, see [CL] **cluster utility** for other cluster commands, including cluster list, that help you manage your analyses. The cluster notes command is modeled after Stata's notes command (see [R] **notes**), but realize that they are different systems and do not interact.

▷ Example

We illustrate the cluster notes command starting with three cluster analyses that have already been performed. The cluster dir command shows us the names of all the existing cluster analyses; see [CL] **cluster utility**.

```
. cluster dir
sngeuc
sngabs
kmn3abs
. cluster note sngabs : I used single linkage with absolute-value distance
. cluster note sngeuc : Euclidean distance and single linkage
. cluster note kmn3abs : This has the kmeans cluster results for 3 groups
```

```
. cluster notes
sngeuc
     notes:   1. Euclidean distance and single linkage

sngabs
     notes:   1. I used single linkage with absolute-value distance

kmn3abs
     notes:   1. This has the kmeans cluster results for 3 groups
```

After adding a note to each of the three cluster analyses, we used the `cluster notes` command without arguments to list all the notes for all the cluster analyses.

The * and ? characters may be used when referring to cluster names; see [U] **14.2 Abbreviation rules**.

```
. cluster note k* : Verify that observation 5 is correct.  I am suspicious that
> there was a typographical error or instrument failure in recording the
> information.
. cluster notes kmn3abs
kmn3abs
     notes:   1. This has the kmeans cluster results for 3 groups
              2. Verify that observation 5 is correct. I am suspicious that
                 there was a typographical error or instrument failure in
                 recording the information.
```

`cluster notes` expanded k* to kmn3abs the only cluster name that begins with a k. Notes that extend to multiple lines are automatically wrapped when displayed. When entering long notes you just continue to type until your note is finished. Pressing return signals that you are done with that note.

After examining the dendrogram (see [CL] **cluster dendrogram**) for the sngeuc single linkage cluster analysis and seeing one small group of data that split off from the main body of data at a very large distance, you investigate further and find data problems. You decide to add some notes to the sngeuc analysis.

```
. cluster note *euc : All of Sam's data look wrong to me.
. cluster note *euc : I think Sam should be fired.
. cluster notes sng?*
sngeuc
     notes:   1. Euclidean distance and single linkage
              2. All of Sam's data look wrong to me.
              3. I think Sam should be fired.

sngabs
     notes:   1. I used single linkage with absolute-value distance
```

Sam, one of the lab technicians, who happens to be the owner's nephew and is paid more than you, really messed up. After adding these notes, you get second thoughts about keeping the notes attached to the cluster analysis (and the data). You decide you really want to delete those notes and to add a more politically correct note.

```
. cluster note sngeuc : Ask Jennifer to help Sam re-evaluate his data.
. cluster note sngeuc
sngeuc
     notes:   1. Euclidean distance and single linkage
              2. All of Sam's data looks wrong to me.
              3. I think Sam should be fired.
              4. Ask Jennifer to help Sam re-evaluate his data.
. cluster note drop sngeuc in 2/3
```

```
. cluster notes kmn3abs s*
kmn3abs
     notes:   1. This has the kmeans cluster results for 3 groups
              2. Verify that observation 5 is correct. I am suspicious that
                 there was a typographical error or instrument failure in
                 recording the information.
sngeuc
     notes:   1. Euclidean distance and single linkage
              2. Ask Jennifer to help Sam re-evaluate his data.
sngabs
     notes:   1. I used single linkage with absolute-value distance
```

Just for illustration purposes the new note was added before deleting the two offending notes. cluster notes drop can take an in argument followed by a list of note numbers. The numbers correspond to those shown in the listing provided by the cluster notes command. After the deletions, the note numbers are reassigned to remove gaps. So, sngeuc note 4 becomes note 2 after the deletion of notes 2 and 3 as shown above.

Without an in argument, the cluster notes drop command drops all notes associated with the named cluster.

◁

Remember that the cluster notes are stored with the data, and, as with other updates you make to the data, the additions and deletions are not permanent until you save the data; see [R] **save**.

❏ Technical Note

Programmers can access the notes (and all the other cluster attributes) using the cluster query command; see [CL] **cluster programming utilities**.

❏

Also See

Complementary:	[CL] **cluster programming utilities**, [CL] **cluster utility**, [R] **save**
Related:	[R] **notes**
Background:	[CL] **cluster**

Title

cluster programming subroutines — Add cluster analysis routines

Description

This entry describes how to extend Stata's `cluster` command; see [CL] **cluster**. Programmers can add subcommands to `cluster`; add functions to `cluster generate` (see [CL] **cluster generate**); add stopping rules to `cluster stop` (see [CL] **cluster stop**); and set up an alternate command to be executed when `cluster dendrogram` is called (see [CL] **cluster dendrogram**).

The `cluster` command also provides utilities for programmers; see [CL] **cluster programming utilities** to learn more.

Remarks

Remarks are presented under the headings

Adding a cluster subroutine
Adding a cluster generate function
Adding a cluster stopping rule
Applying an alternate cluster dendrogram routine

Adding a cluster subroutine

You add a `cluster` subroutine by creating a Stata program with the name `cluster_`*subcmdname*. For example, to add the subcommand xyz to `cluster`, create `cluster_xyz.ado`. Users could then execute the xyz subcommand with

`cluster xyz ...`

Everything entered on the command line following `cluster xyz` is passed to the `cluster_xyz` command.

You can add new clustering methods, new cluster management tools, and new post-clustering programs. The `cluster` command has subcommands available that are helpful to cluster analysis programmers; see [CL] **cluster programming utilities**.

▷ Example

We demonstrate the addition of a `cluster` subroutine by writing a simple post-cluster analysis routine that provides a cross-tabulation of two cluster analysis grouping variables. The syntax of the new command will be

cluster mycrosstab *cname1 cname2* [, *tabulate_options*]

Here is the program:

```
program cluster_mycrosstab
        gettoken clname1 0 : 0 , parse(" ,")
        gettoken clname2 rest : 0 , parse(" ,")
        cluster query `clname1'
        local groupvar1 `r(groupvar)'
        cluster query `clname2'
        local groupvar2 `r(groupvar)'
        tabulate `groupvar1' `groupvar2' `rest'
end
```

71

See [P] **gettoken** for information on the gettoken command, and see [R] **tabulate** for information on the tabulate command. The cluster query command is one of the cluster programming utilities that is documented in [CL] **cluster programming utilities**.

We can demonstrate cluster mycrosstab in action. This example starts with two cluster analyses, cl1 and cl2, that have already been performed. The dissimilarity measure and the variables included in the two cluster analyses differ. We want to see how closely the two cluster analyses match.

```
. cluster list, type method dissim var
cl2  (type: partition,  method: kmeans,  dissimilarity: L(1.5))
      vars: gvar2 (group variable)

cl1  (type: partition,  method: kmeans,  dissimilarity: L1)
      vars: cl1gvar (group variable)

. cluster mycrosstab cl1 cl2, chi2
```

cl1gvar	1	gvar2 2	3	4	5	Total
1	0	0	10	0	0	10
2	1	4	0	5	6	16
3	8	0	0	4	1	13
4	9	0	8	4	0	21
5	0	8	0	0	6	14
Total	18	12	18	13	13	74

```
       Pearson chi2(16) =  98.4708   Pr = 0.000
```

The chi2 option was included to demonstrate that we were able to exploit the existing options of tabulate with very little programming effort. We just pass along to tabulate any of the extra arguments received by cluster_mycrosstab.

◁

Adding a cluster generate function

Programmers can add functions to the cluster generate command; see [CL] **cluster generate**. This is accomplished by creating a command called clusgen_*name*. For example, if I wanted to add a function called abc() to cluster generate, I would create clusgen_abc.ado. Users could then execute

    cluster generate *newvarname* = abc(...) ...

Everything entered on the command line following cluster generate is passed to clusgen_abc.

▷ Example

Here is the beginning of a clusgen_abc program that expects an integer argument and that has one option called name(*clname*) that gives the name of the cluster. If name() is not specified, it defaults to the most recently performed cluster analysis. We will assume, for illustration purposes, that the cluster analysis must be hierarchical, and will check for this in the clusgen_abc program.

```
program clusgen_abc
        // we use gettoken to work our way through the parsing
        gettoken newvar 0 : 0 , parse(" =")
        gettoken temp 0 : 0 , parse(" =")
        if '"'temp'"' != "=" {
                error 198
        }
        gettoken temp 0 : 0 , parse(" (")
        if '"'temp'"' != "abc" {
                error 198
        }
        gettoken funcarg 0 : 0 , parse(" (") match(temp)
        if '"'temp'"' != "(" {
                error 198
        }

        // funcarg holds the integer argument to abc()
        confirm integer number 'funcarg'

        // we can now use syntax to parse the option
        syntax [, Name(str) ]

        // cluster query will give us the list of cluster names
        if '"'name'"' == "" {
                cluster query
                local clnames 'r(names)'
                if "'clnames'" == "" {
                        di as err "no cluster solutions defined"
                        exit 198
                }
                // first name in the list is the latest clustering
                local name : word 1 of 'clnames'
        }

        // cluster query followed by name will tell us the type
        cluster query 'name'
        if "'r(type)'" != "hierarchical" {
                di as err "only allowed with hierarchical clustering"
                exit 198
        }
        /*
           you would now pull more information from the call of
                        cluster query 'name'
           and do your computations and generate 'newvar'
        */
        ...

end
```

See [CL] **cluster programming utilities** for details on the `cluster query` command.

◁

Adding a cluster stopping rule

Programmers can add stopping rules to the `rules()` option of the `cluster stop` command; see
[CL] **cluster stop**. This is done by creating a Stata program with the name `clstop_name`. For example,
to add a stopping rule named `mystop` so that `cluster stop` would now have a `rules(mystop)`
option, create `clstop_mystop.ado` defining the `clstop_mystop` program. Users could then execute

```
cluster stop [clname] , rule(mystop) ...
```

The `clstop_mystop` program is passed the cluster name (*clname*) provided by the user (or the name of the current cluster result if not specified), followed by a comma and all the options entered by the user except for the `rule(mystop)` option.

▷ Example

We demonstrate the addition of a `rule(stepsize)` option to `cluster stop`. This option implements the simple step-size stopping rule mentioned in Milligan and Cooper (1985). The step-size stopping rule computes the difference in fusion values between levels in a hierarchical cluster analysis. (A fusion value is the similarity or dissimilarity measure at which clusters are fused or split in the hierarchical cluster structure.) Large values of the step-size stopping rule indicate groupings with more distinct cluster structure.

People who examine cluster dendrograms (see [CL] **cluster dendrogram**) to visually determine the number of clusters are in essence using a visual approximation to the step-size stopping rule.

Here is the `clstop_stepsize` program:

```
program clstop_stepsize, sortpreserve rclass
        syntax anything(name=clname) [, Depth(integer -1) ]

        cluster query `clname'
        if "`r(type)'" != "hierarchical" {
                di as error ///
                    "rule(stepsize) only allowed with hierarchical clustering"
                exit 198
        }
        if "`r(pseudo_heightvar)'" != "" {
                di as error "dendrogram reversals encountered"
                exit 198
        }
        local hgtvar `r(heightvar)'
        if `""`r(similarity)'""' != "" {
                sort `hgtvar'
                local negsign "-"
        }
        else if `""`r(dissimilarity)'""' != "" {
                gsort -`hgtvar'
        }
        else {
                di as error "dissimilarity or similarity not set"
                exit 198
        }

        quietly count if !missing(`hgtvar')
        local depth = cond(`depth'<=1, r(N), min(`depth',r(N)))

        tempvar diff
        qui gen double `diff'=`negsign'(`hgtvar'-`hgtvar'[_n+1]) if _n<`depth'

        di
        di as txt "Depth" _col(10) "Stepsize"
        di as txt "{hline 17}"
        forvalues i = 1/`= `depth'-1' {
                local j = `i' + 1
                di as res `j' _col(10) %8.0g `diff'[`i']
                return scalar stepsize_`j' = `diff'[`i']
        }
        return local rule "stepsize"
end
```

See [P] **syntax** for information on the `syntax` command, [P] **forvalues** for information on the `forvalues` looping command, and [P] **macro** for information on the '= ... ' macro function. The `cluster query` command is one of the cluster programming utilities that is documented in [CL] **cluster programming utilities**.

With this program, users can obtain the step-size stopping rule. We demonstrate using the average linkage hierarchical cluster analysis found in the second example of [CL] **cluster averagelinkage**. The dataset contains 30 observations on 60 binary variables. The simple matching coefficient is used as the similarity measure in the average linkage clustering.

```
. cluster a a1-a60, match name(alink)

. cluster stop alink, rule(stepsize) depth(15)
```

Depth	Stepsize
2	.065167
3	.187333
4	.00625
5	.007639
6	.002778
7	.005952
8	.002381
9	.008333
10	.005556
11	.002778
12	0
13	0
14	.006667
15	.01

In the `clstop_stepsize` program, we included a `depth()` option. `cluster stop`, when called with the new `rule(stepsize)` option, can also have the `depth()` option. In this case, we asked to stop at a depth of 15.

The largest step-size, .187, happens at the three-group level of the hierarchy. The number .187 is the difference in the matching coefficient when two groups are formed versus when three groups are formed in this hierarchical cluster analysis.

The `clstop_stepsize` program could be enhanced by using a prettier output table format. An option could also be added that saves the results to a matrix.

◁

Applying an alternate cluster dendrogram routine

Programmers can change the behavior of the `cluster dendrogram` command (alias `cluster tree`); see [CL] **cluster dendrogram**. This is accomplished by using the `other()` option of the `cluster set` command (see [CL] **cluster programming utilities**), with a *tag* of `treeprogram` and with *text* giving the name of the command to be used in place of the standard Stata program for `cluster dendrogram`. For example, if I had created a new hierarchical cluster analysis method for Stata that needed a different algorithm for producing dendrograms, I would use the command

```
cluster set clname, other(treeprogram progname)
```

to set *progname* as the program to be executed when `cluster dendrogram` is called.

▷ Example

If I am creating a new hierarchical cluster analysis method called `myclus`, I would create a program called `cluster_myclus` (see the discussion at the beginning of the *Remarks* section). If `myclus` needed a different dendrogram routine from the standard one used within Stata, I would include the following line inside `cluster_myclus.ado` at the point where I set the cluster attributes.

```
cluster set 'clname', other(treeprogram myclustree)
```

I would then create a program called `myclustree` in a file called `myclustree.ado` that implements the particular dendrogram program needed by `myclus`.

◁

References

Milligan, G. W. and M. C. Cooper. 1985. An examination of procedures for determining the number of clusters in a dataset. *Psychometrika* 50: 159–179.

Also See

Background: [CL] **cluster**, [CL] **cluster programming utilities**

Title

> **cluster programming utilities** — Cluster analysis programming utilities

Syntax

cluster query [*clname*]

cluster set [*clname*] [, <u>add</u>name <u>type</u>(*type*) <u>method</u>(*method*)

[<u>s</u>imilarity(*measure*) | <u>d</u>issimilarity(*measure*)] <u>var</u>(*tag varname*)

<u>char</u>(*tag charname*) <u>o</u>ther(*tag text*) <u>note</u>(*text*)]

cluster <u>del</u>ete *clname* [, zap <u>del</u>name type <u>m</u>ethod <u>d</u>issimilarity <u>s</u>imilarity

<u>note</u>s(*numlist*) <u>alln</u>otes <u>var</u>(*tag*) <u>allv</u>ars <u>varz</u>ap(*tag*) <u>allvarzap</u> <u>char</u>(*tag*)

<u>allc</u>hars <u>charz</u>ap(*tag*) <u>allcharzap</u> <u>o</u>ther(*tag*) <u>allo</u>thers]

cluster <u>parse</u>distance *measure*

cluster measures *varlist* [if *exp*] [in *range*] , <u>compare</u>(*numlist*)

<u>g</u>enerate(*newvarlist*) [*measure* <u>propv</u>ars <u>propc</u>ompares]

Description

The cluster query, cluster set, cluster delete, cluster parsedistance, and cluster measures commands provide tools for programmers to add their own cluster analysis subroutines to Stata's cluster command; see [CL] **cluster** and [CL] **cluster programming subroutines**. These commands make it possible for the new command to take advantage of Stata's cluster management facilities.

cluster query provides a way to obtain the various attributes of a cluster analysis in Stata. If *clname* is omitted, cluster query returns in r(names) a list of the names of all currently defined cluster analyses. If *clname* is provided, then the various attributes of the named cluster analysis are returned in r(). These attributes include the type, method, (dis)similarity used, created variable names, notes, and any other information attached to the cluster analysis.

cluster set allows you to set the various attributes that define a cluster analysis in Stata. This includes giving a name to your cluster results and adding that name to the master list of currently defined cluster results. With cluster set, you can provide information on the type, method, and (dis)similarity measure of your cluster analysis results. You can associate variables and Stata characteristics (see [P] **char**) with your cluster analysis. cluster set also allows you to add notes and other named fields to your cluster analysis result. These items become part of the dataset and are saved with the data.

cluster delete allows you to delete attributes from a cluster analysis in Stata. This command is the inverse of cluster set.

cluster parsedistance takes the similarity or dissimilarity *measure* name and checks it against the list of those provided by Stata, taking account of allowed minimal abbreviations and aliases. Aliases are resolved (for instance, Euclidean is changed into the equivalent L2).

cluster measures computes the similarity or dissimilarity *measure* between the observations listed in the compare() option and the observations included based on the if and in conditions, and places the results in the variables specified by the generate() option.

Stata also provides a method for programmers to extend the cluster command by providing subcommands. This is discussed in [CL] **cluster programming subroutines**.

Options

Options for cluster set

addname adds *clname* to the master list of currently defined cluster analyses. When *clname* is not specified, the addname option is mandatory, and in this case, cluster set automatically finds a cluster name that is not currently in use and uses this as the cluster name. cluster set returns the name of the cluster in r(name). If addname is not specified, the *clname* must have previously been added to the master list (for instance, through a previous call to cluster set).

type(*type*) sets the cluster type for *clname*. type(hierarchical) indicates that the cluster analysis is some kind of hierarchical clustering, and type(partition) indicates that it is a partition style of clustering. You are not restricted to these types. For instance, you might program some kind of fuzzy partition clustering analysis, and could, in that case, use type(fuzzy).

method(*method*) sets the name of the clustering method for the cluster analysis. For instance, Stata uses method(kmeans) to indicate a kmeans cluster analysis, and uses method(single) to indicate single linkage cluster analysis. You are not restricted to the names currently employed within Stata.

dissimilarity(*measure*) and similarity(*measure*) set the name of the dissimilarity or similarity measure used for the cluster analysis. For example, Stata uses dissimilarity(L2) to indicate the L2 or Euclidean distance. You are not restricted to the names currently employed within Stata. See [CL] **cluster** for a listing and discussion of (dis)similarity measures.

var(*tag varname*) sets a marker called *tag* in the cluster analysis that points to the variable *varname*. For instance, Stata uses var(group *varname*) to set a grouping variable from a kmeans cluster analysis. With single linkage clustering, Stata uses var(id *idvarname*), var(order *ordervarname*), and var(height *hgtvarname*) to set the id, order, and height variables that define the cluster analysis result. You are not restricted to the names currently employed within Stata. Up to ten var() options may be specified with a single cluster set command.

char(*tag charname*) sets a marker called *tag* in the cluster analysis that points to the Stata characteristic named *charname*; see [P] **char**. This can be either an _dta[] dataset characteristic or a variable characteristic. Up to ten char() options may be specified with a single cluster set command.

other(*tag text*) sets a marker called *tag* in the cluster analysis with *text* attached to the tag marker. Stata uses other(k #) to indicate that k (the number of groups) was # in a kmeans cluster analysis. You are not restricted to the names currently employed within Stata. Up to ten other() options may be specified with a single cluster set command.

note(*text*) adds a note to the *clname* cluster analysis. The cluster notes command (see [CL] **cluster notes**) is the command for users to add, delete, or view cluster notes. The cluster notes command uses the note() option of cluster set to actually add a note to a cluster analysis. Up to ten note() options may be specified with a single cluster set command.

Options for cluster delete

zap deletes everything possible for cluster analysis *clname*. It is the same as specifying the delname, type, method, dissimilarity, similarity, allnotes, allcharzap, allothers, and
allvarzap options.

delname removes *clname* from the master list of current cluster analyses. This option does not touch
the various pieces that make up the cluster analysis. To remove them, use the other options of
cluster delete.

type deletes the cluster type entry from *clname*.

method deletes the cluster method entry from *clname*.

dissimilarity and similarity delete the dissimilarity and similarity entries, respectively, from
clname.

notes(*numlist*) deletes the specified numbered notes from *clname*. The numbering corresponds to the
returned results from the cluster query *clname* command. The cluster notes drop command
(see [CL] **cluster notes**) is the command for users to drop a cluster note. It, in turn, calls cluster
delete, using the notes() option to drop the notes.

allnotes removes all notes from the *clname* cluster analysis.

var(*tag*) removes from *clname* the entry labeled *tag* that points to a variable. This option does not
delete the variable.

allvars removes all the entries pointing to variables for *clname*. This option does not delete the
corresponding variables.

varzap(*tag*) is the same as var(), with the addition of actually deleting the referenced variable.

allvarzap is the same as allvars, with the addition of actually deleting the variables.

char(*tag*) removes from *clname* the entry labeled *tag* that points to a Stata characteristic (see
[P] **char**). This option does not delete the characteristic.

allchars removes all the entries pointing to Stata characteristics for *clname*. This option does not
delete the characteristics.

charzap(*tag*) is the same as char(), with the addition of actually deleting the characteristic.

allcharzap is the same as allchars, with the addition of actually deleting the characteristics.

other(*tag*) deletes from *clname* the *tag* entry and its associated text, which were set using the
other() option of the cluster set command.

allothers deletes all entries from *clname* that have been set using the other() option of the
cluster set command.

Options for cluster measures

compare(*numlist*) is required, and specifies the observations to use as the comparison observations.
Each of these observations will be compared with the observations implied by the if and in
conditions using the specified (dis)similarity *measure*. The results are stored in the corresponding
new variable from the generate() option. There must be the same number of elements in *numlist*
as variable names in the generate() option.

generate(*newvarlist*) is required, and specifies the names of the variables to be created. There must
be as many elements in *newvarlist* as numbers specified in the compare() option.

measure is one of the similarity or dissimilarity measures allowed by Stata. The default is L2 (meaning Euclidean distance). A list of allowed *measures* and their minimal abbreviations can be found in [CL] **cluster**.

propvars is for use with binary measures. It indicates that the observations implied by the if and in conditions are to be interpreted as proportions of binary observations. The default action with binary measures treats all nonzero values as one (excluding missing values). With propvars, the values are confirmed to be between zero and one inclusive. See [CL] **cluster** for a discussion of the use of proportions with binary measures.

propcompares is for use with binary measures. It indicates that the comparison observations (those specified in the compare() option) are to be interpreted as proportions of binary observations. The default action with binary measures treats all nonzero values as one (excluding missing values). With propcompares, the values are confirmed to be between zero and one inclusive. See [CL] **cluster** for a discussion of the use of proportions with binary measures.

Remarks

▷ Example

Programmers can determine which cluster solutions currently exist by using the cluster query command without specifying a cluster name. It returns the names of all currently defined clusters.

```
. cluster dir
grpk7L1
grpk6L1
grpk7L2
grpk6L2
. cluster query
. return list
macros:
          r(names) : "grpk7L1 grpk6L1 grpk7L2 grpk6L2"
```

In this case, there are four cluster solutions. A programmer can further process the r(names) returned macro. For example, if we want to determine which current cluster solutions used kmeans clustering, we would loop through these four cluster solution names, and for each one call cluster query to determine its properties.

```
. local clusnames 'r(names)'
. foreach cname of local clusnames {
  2.          cluster query 'cname'
  3.          if "'r(method)'" == "kmeans" {
  4.                  local kmeancls 'kmeancls' 'cname'
  5.          }
  6. }
. di "{tab}Cluster analyses using kmeans: 'kmeancls'"
        Cluster analyses using kmeans: grpk7L2 grpk6L2
```

In this case, we examined r(method), which records the name of the cluster analysis method. Two of the four cluster solutions used kmeans.

◁

▷ Example

We interactively demonstrate `cluster set`, `cluster delete`, and `cluster query`, though in practice these would be used within a program.

First, we add the name `myclus` to the master list of cluster analyses, and at the same time set the type, method, and similarity.

```
. cluster set myclus, addname type(madeup) method(fake) similarity(who knows)
. cluster query
. return list
macros:
            r(names) : "myclus grpk7L1 grpk6L1 grpk7L2 grpk6L2"
. cluster query myclus
. return list
macros:
             r(name) : "myclus"
       r(similarity) : "who knows"
           r(method) : "fake"
             r(type) : "madeup"
```

`cluster query` shows that `myclus` was successfully added to the master list of cluster analyses, and that the attributes that were `cluster set` can also be obtained.

Now we add a reference to a variable. We will use the word `group` as the *tag* for a variable `mygrpvar`. Additionally, we add another item called `xyz` and associate some text with the `xyz` item.

```
. cluster set myclus, var(group mygrpvar) other(xyz some important info)
. cluster query myclus
. return list
macros:
             r(name) : "myclus"
          r(o1_val) : "some important info"
          r(o1_tag) : "xyz"
        r(groupvar) : "mygrpvar"
         r(v1_name) : "mygrpvar"
          r(v1_tag) : "group"
       r(similarity) : "who knows"
           r(method) : "fake"
             r(type) : "madeup"
```

The `cluster query` command returned the `mygrpvar` information in two ways. The first way is with `r(v#_tag)` and `r(v#_name)`. In this case, there is only one variable associated with `myclus`, so we have `r(v1_tag)` and `r(v1_name)`. This allows the programmer to loop over all the saved variable names without knowing beforehand what the *tag*s might be or how many there are. You could loop as follows:

```
local i 1
while "`r(v`i'_tag)'" != "" {
        ...
        local ++i
}
```

The second way the variable information is returned is in an `r()` result with the *tag* name appended by var, `r(tagvar)`. In our example, this is `r(groupvar)`. This second method is convenient when, as the programmer, you know exactly which *varname* information you are seeking.

The same logic applies with regards to characteristic attributes that are `cluster set`.

Now we continue with our interactive example:

```
. cluster delete myclus, method var(group)
. cluster set myclus, note(a note) note(another note) note(a third note)
. cluster query myclus
. return list

macros:
                 r(name) : "myclus"
                r(note3) : "a third note"
                r(note2) : "another note"
                r(note1) : "a note"
               r(o1_val) : "some important info"
               r(o1_tag) : "xyz"
           r(similarity) : "who knows"
                 r(type) : "madeup"
```

We used `cluster delete` to remove the method and the `group` variable we had associated with `myclus`. Three notes were then simultaneously added using the `note()` option of `cluster set`. In practice, users will use the `cluster notes` command (see [CL] **cluster notes**) to add and delete cluster notes. The `cluster notes` command is implemented using the `cluster set` and `cluster delete` programming commands.

We finish our interactive demonstration of these commands by deleting some more attributes from `myclus` and then completely eliminating `myclus`. In practice, users would remove a cluster analysis with the `cluster drop` command (see [CL] **cluster utility**), which is implemented using the `zap` option of the `cluster delete` command.

```
. cluster delete myclus, allnotes similarity
. cluster query myclus
. return list

macros:
                 r(name) : "myclus"
               r(o1_val) : "some important info"
               r(o1_tag) : "xyz"
                 r(type) : "madeup"
. cluster delete myclus, zap
. cluster query
. return list

macros:
                r(names) : "grpk7L1 grpk6L1 grpk7L2 grpk6L2"
```

The cluster attributes that are `cluster set` become a part of the dataset. They are saved with the dataset when it is saved, and are available again when the dataset is used; see [R] **save**.

◁

❑ Technical Note

You may wonder how Stata's cluster analysis data structures are implemented. Stata data characteristics (see [P] **char**) hold the information. The details of the implementation are not important, and in fact, we encourage you to use the `set`, `delete`, and `query` subcommands to access the cluster attributes. This way, if we ever decide to change the underlying implementation, you will be protected through Stata's version control feature.

❑

▷ Example

The `cluster parsedistance` programming command takes as an argument the name of a similarity or dissimilarity measure. Stata then checks this name against those implemented within Stata (and available to you through the `cluster measures` command). Uppercase or lowercase letters are allowed, and minimal abbreviations are checked. Some of the measures have aliases. These are resolved so that a standard measure name is returned. We demonstrate the `cluster parsedistance` command interactively:

```
. cluster parsedistance max
. sreturn list

macros:
            s(drange) : "0 ."
            s(dtype) : "dissimilarity"
             s(dist) : "Linfinity"
. cluster parsedistance Eucl
. sreturn list

macros:
            s(drange) : "0 ."
            s(dtype) : "dissimilarity"
             s(dist) : "L2"
. cluster parsedistance correl
. sreturn list

macros:
            s(drange) : "1 -1"
            s(dtype) : "similarity"
             s(dist) : "correlation"
. cluster parsedistance jacc
. sreturn list

macros:
            s(drange) : "1 0"
           s(binary) : "binary"
            s(dtype) : "similarity"
             s(dist) : "Jaccard"
```

`cluster parsedistance` returns `s(dtype)` as either `similarity` or `dissimilarity`, and returns `s(dist)` as the standard Stata name for the (dis)similarity. `s(drange)` gives the range of the measure (most similar to most dissimilar). If the measure is designed for binary variables, then `s(binary)` is returned with the word `binary`, as seen above.

See [CL] **cluster** for a listing of the similarity and dissimilarity measures and their properties.

◁

▷ Example

`cluster measures` computes the similarity or dissimilarity measure between each comparison observation and the observations implied by the `if` and `in` conditions (or all the data if no `if` or `in` conditions are specified).

We demonstrate using the auto dataset:

```
. use http://www.stata-press.com/data/r8/auto
. cluster measures turn trunk gear_ratio in 1/10, compare(3 11) gen(z3 z11) L1
. format z* %8.2f
```

```
. list turn trunk gear_ratio z3 z11 in 1/11
```

	turn	trunk	gear_r~o	z3	z11
1.	40	11	3.58	6.50	14.30
2.	40	11	2.53	6.55	13.25
3.	35	12	3.08	0.00	17.80
4.	40	16	2.93	9.15	8.65
5.	43	20	2.41	16.67	1.13
6.	43	21	2.73	17.35	2.45
7.	34	10	2.87	3.21	20.59
8.	42	16	2.93	11.15	6.65
9.	43	17	2.93	13.15	4.65
10.	42	13	3.08	8.00	9.80
11.	44	20	2.28	.	.

Using the three variables turn, trunk, and gear_ratio, we computed the L1 (or absolute value) distance between the third observation and the first ten observations, and placed the results in the variable z3. The distance between the eleventh observation and the first ten was placed in variable z11.

There are many measures designed for binary data. Below we illustrate cluster measures with the matching coefficient binary similarity measure. We have eight observations on ten binary variables, and we will compute the matching similarity measure between the last three observations and all eight observations.

```
. cluster measures x1-x10, compare(6/8) gen(z6 z7 z8) matching
. format z* %4.2f
. list
```

	x1	x2	x3	x4	x5	x6	x7	x8	x9	x10	z6	z7	z8
1.	1	0	0	0	1	1	0	0	1	1	0.60	0.80	0.40
2.	1	1	1	0	0	1	0	1	1	0	0.70	0.30	0.70
3.	0	0	1	0	0	0	1	0	0	1	0.60	0.40	0.20
4.	1	1	1	1	0	0	0	1	1	1	0.40	0.40	0.60
5.	0	1	0	1	1	0	1	0	0	1	0.20	0.60	0.40
6.	1	0	1	0	0	1	0	0	0	0	1.00	0.40	0.60
7.	0	0	0	1	1	1	0	0	1	1	0.40	1.00	0.40
8.	1	1	0	1	0	1	0	1	0	0	0.60	0.40	1.00

Stata will treat all nonzero observations as being one (except missing values, which are treated as missing values) when computing these binary measures.

When the similarity measure between binary observations and the means of groups of binary observations is needed, the propvars and propcompares options of cluster measures provide the solution. The mean of binary observations is a proportion. The value 0.2 would indicate that 20 percent of the values were one and 80 percent were zero for the group. See [CL] cluster for a discussion of binary measures. The propvars option indicates that the main body of observations should be interpreted as proportions. The propcompares option indicates that the comparison observations should be treated as proportions.

We compare ten binary observations on five variables to two observations holding proportions by using the propcompares option:

. cluster measures a* in 1/10, compare(11 12) gen(c1 c2) matching propcompare
. list

	a1	a2	a3	a4	a5	c1	c2
1.	1	1	1	0	1	.6	.56
2.	0	0	1	1	1	.36	.8
3.	1	0	1	0	0	.76	.56
4.	1	1	0	1	1	.36	.44
5.	1	0	0	0	0	.68	.4
6.	0	0	1	1	1	.36	.8
7.	1	0	1	0	1	.64	.76
8.	1	0	0	0	1	.56	.6
9.	0	1	1	1	1	.32	.6
10.	1	1	1	1	1	.44	.6
11.	.8	.4	.7	.1	.2	.	.
12.	.5	0	.9	.6	1	.	.

◁

Saved Results

cluster query with no arguments saves in r():

Macros
    r(names)   cluster solution names

cluster query with an argument saves in r():

Macros
r(name)	cluster name
r(type)	type of cluster analysis
r(method)	cluster analysis method
r(similarity)	similarity measure name
r(dissimilarity)	dissimilarity measure name
r(note#)	cluster note number #
r(v#_tag)	variable tag number #
r(v#_name)	varname associated with r(v#_tag)
r(*tag*var)	varname associated with *tag*
r(c#_tag)	characteristic tag number #
r(c#_name)	characteristic name associated with r(c#_tag)
r(c#_val)	characteristic value associated with r(c#_tag)
r(*tag*char)	characteristic name associated with *tag*
r(o#_tag)	other tag number #
r(o#_val)	other value associated with r(o#_tag)

cluster set saves in r():

Macros
    r(name)   cluster name

`cluster parsedistance` saves in `s()`:

> Macros
> | | |
> |---|---|
> | s(dist) | (dis)similarity measure name |
> | s(darg) | argument of (dis)similarities that take them, such as L(#) |
> | s(dtype) | the word similarity or dissimilarity |
> | s(drange) | range of measure (most similar to most dissimilar) |
> | s(binary) | the word binary if the measure is for binary observations |

`cluster measures` saves in `r()`:

> Macros
> | | |
> |---|---|
> | r(generate) | variable names from the generate() option |
> | r(compare) | observation numbers from the compare() option |
> | r(dtype) | the word similarity or dissimilarity |
> | r(distance) | the name of the (dis)similarity measure |
> | r(binary) | the word binary if the measure is for binary observations |

Also See

Background: [CL] **cluster**, [CL] **cluster programming subroutines**

Title

> **cluster singlelinkage** — Single linkage cluster analysis

Syntax

cluster <u>s</u>inglelinkage [*varlist*] [if *exp*] [in *range*] [, <u>n</u>ame(*clname*)

    distance_option <u>gen</u>erate(*stub*)]

Description

The cluster singlelinkage command performs hierarchical agglomerative single linkage cluster analysis, which is also known as the nearest-neighbor technique. See [CL] **cluster** for a general discussion of cluster analysis and for a description of the other cluster commands. The cluster dendrogram command (see [CL] **cluster dendrogram**) will display the resulting dendrogram, the cluster stop command (see [CL] **cluster stop**) will help in determining the number of groups, and the cluster generate command (see [CL] **cluster generate**) will produce grouping variables.

Options

name(*clname*) specifies the name to attach to the resulting cluster analysis. If name() is not specified, Stata finds an available cluster name, displays it for your reference, and then attaches the name to your cluster analysis.

distance_option is one of the similarity or dissimilarity measures allowed by Stata. Capitalization of the option does not matter. See [CL] **cluster** for a discussion of these measures.

The available measures designed for continuous data are L2 (synonym <u>Euclid</u>ean), which is the default; L2squared; L1 (synonyms <u>absolute</u>, <u>cityblock</u>, and <u>manhattan</u>); Linfinity (synonym <u>maximum</u>); L(#); <u>Lpower</u>(#); Canberra; <u>corr</u>elation; and angular (synonym <u>angle</u>).

The available measures designed for binary data are <u>matching</u>, <u>Jaccard</u>, <u>Russell</u>, Hamman, Dice, antiDice, Sneath, Rogers, Ochiai, Yule, <u>Anderberg</u>, <u>Kulczynski</u>, Gower2, and Pearson.

generate(*stub*) provides a prefix for the variable names created by cluster singlelinkage. By default, the variable-name prefix will be the name specified in name(). Three variables are created and attached to the cluster analysis results, with the suffixes _id, _ord, and _hgt. Users generally will not need to access these variables directly.

Remarks

An example using the default L2 (Euclidean) distance on continuous data and an example using the matching coefficient on binary data illustrate the cluster singlelinkage command. The data from these two examples are also used in [CL] **cluster averagelinkage**, [CL] **cluster centroidlinkage**, [CL] **cluster completelinkage**, [CL] **cluster medianlinkage**, [CL] **cluster wardslinkage**, and [CL] **cluster waveragelinkage**, so that you can compare different hierarchical clustering methods.

▷ Example ·

As the senior data analyst for a small biotechnology firm, you are given a dataset with 4 chemical laboratory measurements on 50 different samples of a particular plant gathered from the rain forest. The head of the expedition that gathered the samples thinks, based on information from the natives, that an extract from the plant might reduce the negative side effects associated with your company's best-selling nutritional supplement.

While the company chemists and botanists continue exploring the possible uses of the plant, and plan future experiments, the head of product development asks you to look at the preliminary data and to report anything that might be helpful to the researchers.

While all 50 of the plants are supposed to be of the same type, you decide to perform a cluster analysis to see if there are subgroups or anomalies among them. You arbitrarily decide to use single linkage clustering with the default Euclidean distance.

```
. use http://www.stata-press.com/data/r8/labtech
. cluster singlelinkage x1 x2 x3 x4, name(sngeuc)
. cluster list sngeuc
sngeuc  (type: hierarchical,  method: single,  dissimilarity: L2)
      vars: sngeuc_id (id variable)
            sngeuc_ord (order variable)
            sngeuc_hgt (height variable)
     other: range: 0 .
            cmd: cluster singlelinkage x1 x2 x3 x4, name(sngeuc)
            varlist: x1 x2 x3 x4
```

The `cluster singlelinkage` command generated some variables and created a cluster object with the name `sngeuc`, which you supplied as an argument. `cluster list` provides details about the cluster object; see [CL] **cluster utility**.

What you really want to see is the dendrogram for this cluster analysis; see [CL] **cluster dendrogram**.

```
. cluster dendrogram sngeuc, vertlab ylab
```

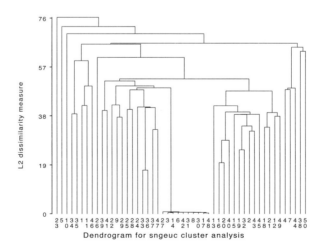

From your experience with looking at dendrograms, two things jump out at you about this cluster analysis. The first is the observations showing up in the middle of the dendrogram that are all very close to each other (very short vertical bars) and are far from any other observations (the long vertical bar connecting them to the rest of the dendrogram). Next you notice that if you ignore those ten observations, the rest of the dendrogram does not indicate strong clustering, as evidenced by the relatively short vertical bars in the upper portion of the dendrogram.

You start to look for clues as to why these ten observations are so peculiar. Looking at scatter plots is usually helpful, and so you examine the matrix of scatter plots.

```
. graph matrix x1 x2 x3 x4
```

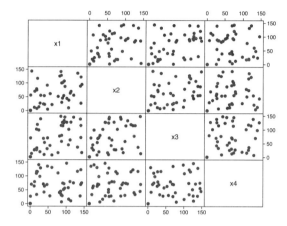

Unfortunately, these scatterplots do not indicate what might be going on.

Suddenly, based on your past experience with the laboratory technicians, you have an idea of what to check next. Because of past data mishaps, the company started the policy of placing within each dataset a variable giving the name of the technician who produced the measurement. You decide to view the dendrogram, using the technician's name as the label instead of the default observation number.

```
. cluster dendrogram sngeuc, vertlab ylab labels(labtech)
```

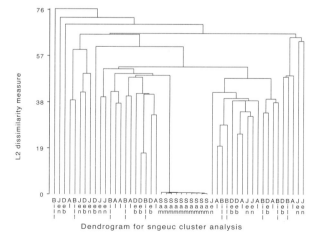

Your suspicions are confirmed. Sam, one of the laboratory technicians, has messed up once again. You list the data and see that all of his observations are between zero and one, while the other four technicians' data range up to about 150, as expected. It looks like Sam forgot, once again, to calibrate his sensor before analyzing his samples. You decide to save a note of your findings with this cluster analysis (see [CL] **cluster notes** for the details) and to send the data back to the laboratory to be fixed.

◁

▷ Example

The sociology professor of your graduate-level class gives, as homework, a dataset containing 30 observations on 60 binary variables, with the assignment to tell him something about the 30 subjects represented by the observations. You feel that this assignment is too vague, but, since your grade depends on it, you get to work trying to figure something out.

Among the analyses you try is the following cluster analysis. You decide to use single linkage clustering with the simple matching binary coefficient, since it is easy to understand. Just for fun, though it makes no difference to you, you specify the generate() option to force the generated variables to have zstub as a prefix. You let Stata pick a name for your cluster analysis by not specifying the name() option.

```
. use http://www.stata-press.com/data/r8/homework
. cluster s a1-a60, matching gen(zstub)
cluster name: _cl_1
. cluster list
_cl_1 (type: hierarchical, method: single, similarity: matching)
      vars: zstub_id (id variable)
            zstub_ord (order variable)
            zstub_hgt (height variable)
     other: range: 1 0
            cmd: cluster singlelinkage a1-a60, matching gen(zstub)
            varlist: a1 a2 a3 a4 a5 a6 a7 a8 a9 a10 a11 a12 a13 a14 a15 a16 a17
                 a18 a19 a20 a21 a22 a23 a24 a25 a26 a27 a28 a29 a30 a31 a32
                 a33 a34 a35 a36 a37 a38 a39 a40 a41 a42 a43 a44 a45 a46 a47
                 a48 a49 a50 a51 a52 a53 a54 a55 a56 a57 a58 a59 a60
```

Stata selected _cl_1 as the cluster name, and created the variables zstub_id, zstub_ord, and zstub_hgt.

You display the dendrogram using the cluster tree command, which is a synonym for cluster dendrogram. Since Stata uses the most recently performed cluster analysis by default, you do not need to type the name.

(Continued on next page)

. cluster tree

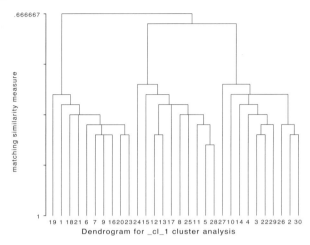

The dendrogram seems to indicate the presence of 3 groups among the 30 observations. You decide that this is probably the structure your teacher wanted you to find, and you begin to write up your report. You want to examine the three groups further, so you use the `cluster generate` command (see [CL] **cluster generate**) to create a grouping variable to make the task easier. You examine various summary statistics and tables for the three groups and finish your report.

After the assignment is turned in, your professor gives you the identical dataset with the addition of one more variable, `truegrp`, which indicates the groupings he feels are in the data. You do a cross-tabulation of the `truegrp` and `grp3`, your grouping variable, to see if you are going to get a good grade on the assignment.

```
. cluster gen grp3 = group(3)
. table grp3 truegrp
```

	truegrp		
grp3	1	2	3
1		10	
2			10
3	10		

Other than the numbers arbitrarily assigned to the three groups, both you and your professor are in complete agreement. You rest easier that night knowing that you may survive one more semester.

◁

Methods and Formulas

[CL] **cluster** discusses hierarchical clustering, and places single linkage clustering in this general framework. It compares the typical behavior of single linkage with complete and average linkage.

Conceptually, hierarchical agglomerative single linkage clustering proceeds as follows. The N observations start out as N separate groups, each of size one. The two closest observations are merged into one group, producing $N - 1$ total groups. The closest two groups are then merged, so that there are $N - 2$ total groups. This process continues until all the observations are merged into one large group. This produces a hierarchy of groupings from one group to N groups. For single linkage clustering, the definition of "closest two groups" is based on the closest observations between the two groups.

Stata's implementation of single linkage clustering is modeled after the algorithm presented in Sibson (1973) and mentioned in Rohlf (1982). This algorithm produces two variables that act as a pointer representation of a dendrogram. To this, Stata adds a third variable used to restore the sort order, as needed, so that the two variables of the pointer representation remain valid. The first variable of the pointer representation gives the order of the observations. The second variable has one less element, and gives the height in the dendrogram at which the adjacent observations in the order-variable join.

See [CL] **cluster** for the details and formulas of the available *distance_option*s, which include (dis)similarity measures for continuous and for binary data.

References

See [CL] **cluster** for references related to cluster analysis, including single linkage clustering. The following references are especially important in terms of the implementation of single linkage clustering:

Day, W. H. E. and H. Edelsbrunner. 1984. Efficient algorithms for agglomerative hierarchical clustering methods. *Journal of Classification* 1: 7–24.

Rohlf, F. J. 1982. Single-link clustering algorithms. In *Handbook of Statistics*, Vol. 2, ed. P. R. Krishnaiah and L. N. Kanal, 267–284. Amsterdam: North Holland Publishing Company.

Sibson, R. 1973. SLINK: An optimally efficient algorithm for the single-link cluster method. *Computer Journal* 16: 30–34.

Also See

Complementary:	[CL] **cluster dendrogram**, [CL] **cluster generate**, [CL] **cluster notes**, [CL] **cluster stop**, [CL] **cluster utility**
Related:	[CL] **cluster averagelinkage**, [CL] **cluster centroidlinkage**, [CL] **cluster completelinkage**, [CL] **cluster medianlinkage**, [CL] **cluster wardslinkage**, [CL] **cluster waveragelinkage**
Background:	[CL] **cluster**

Title

> **cluster stop** — Cluster analysis stopping rules

Syntax

> cluster stop [*clname*] [, <u>ru</u>le(<u>cal</u>inski | duda | *rule_name*) <u>gro</u>ups(*numlist*)
>
> <u>mat</u>rix(*matname*)]

Description

Cluster analysis stopping rules are used to determine the number of clusters. A stopping-rule value (also called an index) is computed for each cluster solution (e.g., at each level of the hierarchy in a hierarchical cluster analysis). Larger values (or smaller, depending on the particular stopping rule) indicate more distinct clustering. See [CL] **cluster** for background information on cluster analysis and on the cluster command.

The cluster stop command currently provides two stopping rules, the Caliński & Harabasz (1974) pseudo-F index and the Duda & Hart (1973) Je(2)/Je(1) index. For both of these rules, larger values indicate more distinct clustering. Presented with the Duda & Hart Je(2)/Je(1) values are pseudo T-squared values. Smaller values of the pseudo T-squared indicate more distinct clustering.

clname specifies the name of the cluster analysis. The default is the latest performed cluster analysis, which can be reset using the cluster use command; see [CL] **cluster utility**.

Additional cluster stop rules may be added; see [CL] **cluster programming subroutines**, which illustrates this by showing a program that adds the step-size stopping rule.

Options

rule(calinski | duda | *rule_name*) indicates the stopping rule. rule(calinski), the default, specifies the Caliński & Harabasz pseudo-F index. rule(duda) specifies the Duda & Hart Je(2)/Je(1) index.

rule(calinski) is allowed for both hierarchical and nonhierarchical cluster analyses. rule(duda) is only allowed with hierarchical cluster analyses.

Additional stopping rules can be added to the cluster stop command; see [CL] **cluster programming subroutines** for details. These additional stopping rules are then obtained using the rule(*rule_name*) option. [CL] **cluster programming subroutines** illustrates the ability to add stopping rules by showing a program that adds a rule(stepsize) option, which implements the simple step-size stopping rule mentioned in Milligan and Cooper (1985).

groups(*numlist*) specifies the cluster groupings for which the stopping rule is to be computed. groups(3/20) indicates that the measure is to be computed for the 3-group solution, the 4-group solution, ..., and the 20-group solution.

With rule(duda), the default is groups(1/15). With rule(calinski) for a hierarchical cluster analysis, the default is groups(2/15). groups(1) is not allowed with rule(calinski) since the measure is not defined for the degenerate 1-group cluster solution. The groups() option is unnecessary (and not allowed) for a nonhierarchical cluster analysis.

If there are ties in the hierarchical cluster analysis structure, some (or possibly all) of the requested stopping-rule solutions may not be computable. `cluster stop` passes over, without comment, the `groups()` for which ties in the hierarchy cause the stopping rule to be undefined.

`matrix(`*matname*`)` saves the results in a matrix named *matname*.

With `rule(calinski)`, the matrix has two columns. The first column gives the number of clusters, and the second column gives the corresponding Caliński & Harabasz pseudo-F stopping-rule index.

With `rule(duda)`, the matrix has three columns. The first column gives the number of clusters. The second column gives the corresponding Duda & Hart Je(2)/Je(1) stopping-rule index. The third column provides the corresponding pseudo T-squared values.

Remarks

Everitt, Landau, and Leese (2001) and Gordon (1999) discuss the problem of determining the number of clusters, and describe several stopping rules, including the Caliński & Harabasz (1974) pseudo-F index and the Duda & Hart (1973) Je(2)/Je(1) index. There are a large number of cluster stopping rules. Milligan and Cooper (1985) provide an evaluation of 30 stopping rules. The Caliński & Harabasz index and the Duda & Hart index were singled out as two of the best rules in their evaluation.

Large values of the Caliński & Harabasz pseudo-F index indicate distinct clustering. The Duda & Hart Je(2)/Je(1) index has an associated pseudo T-squared value. A large value of the Je(2)/Je(1) index and a small value of the pseudo T-squared indicate distinct clustering. See *Methods and Formulas* at the end of this entry for details.

Some stopping rules only work with a hierarchical cluster analysis. The Duda & Hart index is one of them. The Caliński & Harabasz index, however, may be applied to both nonhierarchical and hierarchical cluster analyses.

▷ Example

Previously, you ran kmeans and kmedians cluster analyses on data where you measured the flexibility, speed, and strength of the 80 students in your physical education class; see the first examples of [CL] **cluster kmeans** and [CL] **cluster kmedians**. Your original goal was to split the class into four groups, though you also examined the three- and five-group kmeans and kmedian cluster solutions as possible alternatives.

As described in [CL] **cluster kmedians**, you finally decided that while the four-group solution seemed the best from a clustering standpoint, the five-group solution given by kmedian clustering would work best for your situation.

Now, out of curiosity, you wonder what the Caliński & Harabasz stopping rule shows for the three-, four-, and five-group solutions from your kmedian clustering.

```
. use http://www.stata-press.com/data/r8/physed
. cluster kmed flex speed strength, k(3) name(kmed3) abs start(lastk)
. cluster kmed flex speed strength, k(4) name(kmed4) abs start(kr(11736))
. cluster kmed flex speed strength, k(5) name(kmed5) abs start(prand(8723))
. cluster stop kmed3
```

Number of clusters	Calinski/ Harabasz pseudo-F
3	132.75

```
. cluster stop kmed4
```

Number of clusters	Calinski/ Harabasz pseudo-F
4	337.10

```
. cluster stop kmed5
```

Number of clusters	Calinski/ Harabasz pseudo-F
5	300.45

The four-group solution with a Caliński & Harabasz pseudo-F value of 337.10 is largest, indicating that the four-group solution is the most distinct compared with the three-group and five-group solutions.

The three-group solution has a much lower stopping-rule value of 132.75. The five-group solution, with a value of 300.45, is reasonably close to the four-group solution.

Though you do not think it will change your decision on how to split your class into groups, you are curious to see what a hierarchical cluster analysis might produce. You decide to try an average linkage cluster analysis using the default Euclidean distance; see [CL] **cluster averagelinkage**. You examine the resulting cluster analysis with the `cluster tree` command, which is an easier-to-type alias for the `cluster dendrogram` command; see [CL] **cluster dendrogram**.

```
. cluster averagelink flex speed strength, name(avglnk)
```
```
. cluster tree avglnk, vertlabels ylab
```

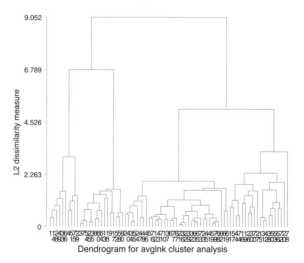

Dendrogram for avglnk cluster analysis

You are curious to see how the four- and five-group solutions from this hierarchical cluster analysis compare with the four- and five-group solutions from the kmedian clustering.

```
. cluster gen avgg = groups(4/5), name(avglnk)
```

```
. table kmed4 avgg4
```

	avgg4			
kmed4	1	2	3	4
1			35	
2		15		
3				20
4	10			

```
. table kmed5 avgg5
```

	avgg5				
kmed5	1	2	3	4	5
1		15			
2				19	1
3			20		
4	10				
5			15		

The four-group solutions are identical except for the numbers used to label the groups. The five-group solutions are different. The kmedian clustering split the 35-member group into subgroups having 20 and 15 members. The average linkage clustering instead split one member off of the 20-member group.

Now you examine the Caliński & Harabasz pseudo-F stopping-rule values associated with the kmedian hierarchical cluster analysis.

```
. cluster stop avglnk, rule(calinski)
```

Number of clusters	Calinski/ Harabasz pseudo-F
2	131.86
3	126.62
4	337.10
5	269.07
6	258.40
7	259.37
8	290.78
9	262.86
10	258.53
11	249.93
12	247.85
13	247.53
14	236.98
15	226.51

Since rule(calinski) is the default, you could have obtained this same table by typing

```
. cluster stop avglnk
```

or, since avglnk was the most recent cluster analysis performed, by typing

```
. cluster stop
```

You did not specify the number of groups to examine from the hierarchical cluster analysis, and so it defaulted to examining up to 15 groups. The highest Caliński & Harabasz pseudo-F is 337.10 for the four-group solution.

What does the Duda & Hart stopping rule produce for this hierarchical cluster analysis?

```
. cluster stop avglnk, rule(duda) groups(1/10)
```

Number of clusters	Je(2)/Je(1)	Duda/Hart pseudo T-squared
1	0.3717	131.86
2	0.1349	147.44
3	0.2283	179.19
4	0.8152	4.08
5	0.2232	27.85
6	0.5530	13.74
7	0.5287	29.42
8	0.6887	3.16
9	0.4888	8.37
10	0.7621	7.80

This time we asked to see the results for one to ten groups. The largest Duda & Hart Je(2)/Je(1) stopping-rule value is 0.8152 corresponding to four groups. The smallest pseudo T-squared is 3.16 for the eight-group solution, but notice that the pseudo T-squared for the four-group solution is also low, with a value of 4.08.

Distinct clustering is characterized by large values of the Caliński & Harabasz pseudo-F, large values of the Duda & Hart Je(2)/Je(1), and small values of the Duda & Hart pseudo T-squared.

The conventional wisdom for deciding the number of groups based on the Duda & Hart stopping-rule table is to find one of the largest Je(2)/Je(1) values that corresponds to a low pseudo T-squared value that has much larger T-squared values next to it. This strategy combined with the results from the Caliński & Harabasz results indicates that the four-group solution is the most distinct from this hierarchical cluster analysis.

Despite all this, you decide to stick with the five-group solution found using kmedian clustering.

◁

❑ Technical Note

There is a good reason that the word "pseudo" appears in "pseudo-F" and "pseudo T-squared". While these index values are based on well-known statistics, any p-values computed from these statistics would not be valid. Remember that cluster analysis searches for structure.

If you were to generate random observations, perform a cluster analysis, compute these stopping-rule statistics, and then follow that with a computation of what would normally be the p-values associated with the statistics, you would almost always end up with significant p-values.

Remember that you would expect, on average, 5 out of every 100 groupings of your random data to show up as significant when you use .05 as your threshold for declaring significance. Cluster analysis methods search for the best groupings, so there is no surprise that p-values show high significance even when none exists.

Examining the stopping-rule index values relative to one another is useful, however, in finding relatively reasonable groupings that may exist in the data.

❑

❏ Technical Note

As mentioned in the *Methods and Formulas* section, ties in the hierarchical cluster structure cause some of the stopping-rule index values to be undefined. Discrete (as opposed to continuous) data tend to cause ties in a hierarchical clustering. The more discrete the data, the more likely it is that ties will occur (and the more of them you will encounter) within a hierarchy.

Even with so-called continuous data, ties in the hierarchical clustering can occur. We say "so called", because most continuous data are truncated or rounded. For instance, miles per gallon, length, weight, etc., which may really be continuous, may be observed and recorded only to the tens, ones, tenths, or hundredths of a unit.

You can have data with no ties in the observations and still have lots of ties in the hierarchy. It is ties in distances (or similarities) between observations and groups of observations that cause the ties in the hierarchy.

Because of this, do not be surprised when some (many) of the stopping-rule values that you request are not presented. Stata has decided not to break the ties arbitrarily, since the stopping-rule values may be very different depending on which split is made.

❏

❏ Technical Note

The stopping rules also become less informative as the number of elements in the groups becomes small; that is, many groups, and each with few observations. We recommend that if you feel a need to examine the stopping-rule values deep within your hierarchical cluster analysis, you do so skeptically.

❏

Saved Results

cluster stop with rule(calinski) saves in r():

Scalars
  r(calinski_#)   Caliński & Harabasz pseudo-F for # groups

Macros
  r(rule)   calinski
  r(label)   C-H pseudo-F
  r(longlabel)   Calinski & Harabasz pseudo-F

cluster stop with rule(duda) saves in r():

Scalars
  r(duda_#)   Duda & Hart Je(2)/Je(1) value for # groups
  r(dudat2_#)   Duda & Hart pseudo T-squared value for # groups

Macros
  r(rule)   duda
  r(label)   D-H Je(2)/Je(1)
  r(longlabel)   Duda & Hart Je(2)/Je(1)
  r(label2)   D-H pseudo T-squared
  r(longlabel2)   Duda & Hart pseudo T-squared

Methods and Formulas

`cluster stop` is implemented as an ado-file.

The Caliński & Harabasz pseudo-F stopping-rule index for g groups and N observations is

$$\frac{\text{trace}(\mathbf{B})/(g-1)}{\text{trace}(\mathbf{W})/(N-g)}$$

where $\mathbf{B}$ is the between-cluster sum of squares and cross-products matrix, and $\mathbf{W}$ is the within-cluster sum of squares and cross-products matrix.

Large values of the Caliński & Harabasz pseudo-F stopping-rule index indicate distinct cluster structure. Small values indicate less clearly defined cluster structure.

The Duda & Hart Je(2)/Je(1) stopping-rule index value is literally Je(2) divided by Je(1). Je(1) is the sum of squared errors within the group that is to be divided. Je(2) is the sum of squared errors in the two resulting subgroups.

Large values of the Duda & Hart pseudo T-squared stopping-rule index indicate distinct cluster structure. Small values indicate less clearly defined cluster structure.

The Duda & Hart Je(2)/Je(1) index requires hierarchical clustering information. It needs to know at each level of the hierarchy which group is to be split and how. The Duda & Hart index is also local since the only information used comes from the group being split. The information in the rest of the groups does not enter the computation.

In comparison, the Caliński & Harabasz rule does not require hierarchical information, and is global since the information from each group is used in the computation.

A pseudo T-squared value is also presented with the Duda & Hart Je(2)/Je(1) index. The relationship is

$$\frac{1}{\text{Je(2)/Je(1)}} = 1 + \frac{T^2}{N_1 + N_2 - 2}$$

where N_1 and N_2 are the numbers of observations in the two subgroups.

Notice that Je(2)/Je(1) will be zero when Je(2) is zero; i.e., when the two subgroups each have no variability. An example of this is when the cluster being split has two distinct values that are being split into singleton subgroups. Je(1) will never be zero because we do not split groups that have no variability. When Je(2)/Je(1) is zero, the pseudo T-squared value is undefined.

Ties in splitting a hierarchical cluster analysis create an ambiguity for the Je(2)/Je(1) measure. For example, to compute the measure for the case of going from five clusters to six, you need to identify the one cluster that will be split. With a tie in the hierarchy, you would instead go from five clusters directly to seven (just as an example). Stata refuses to produce an answer in this situation.

References

Caliński, T. and J. Harabasz. 1974. A dendrite method for cluster analysis. *Communications in Statistics* 3: 1–27.

Duda, R. O. and P. E. Hart. 1973. *Pattern Classification and Scene Analysis.* New York: John Wiley & Sons.

Everitt, B. S., S. Landau, and M. Leese. 2001. *Cluster Analysis.* 4th ed. London: Edward Arnold.

Gordon, A. D. 1999. *Classification.* 2d ed. Boca Raton, FL: CRC Press.

Milligan, G. W. and M. C. Cooper. 1985. An examination of procedures for determining the number of clusters in a dataset. *Psychometrika* 50: 159–179.

Also See

Complementary:	[CL] **cluster averagelinkage**, [CL] **cluster centroidlinkage**, [CL] **cluster completelinkage**, [CL] **cluster kmeans**, [CL] **cluster kmedians**, [CL] **cluster medianlinkage**, [CL] **cluster programming subroutines**, [CL] **cluster singlelinkage**, [CL] **cluster wardslinkage**, [CL] **cluster waveragelinkage**
Related:	[CL] **cluster dendrogram**, [CL] **cluster generate**
Background:	[CL] **cluster**

Title

cluster utility — List, rename, use, and drop cluster analyses

Syntax

```
cluster dir

cluster list [clnamelist] [, all notes type method dissimilarity similarity
    vars chars other ]

cluster drop { clnamelist | _all }

cluster use clname

cluster rename oldclname newclname

cluster renamevar oldvarname newvarname [ , name(clname) ]

cluster renamevar oldstub newstub , prefix [ name(clname) ]
```

Description

These cluster utility commands allow you to view and to manipulate the cluster objects that you have created. See [CL] **cluster** for an overview of cluster analysis and for the available cluster commands. If you desire even more control over your cluster objects, or if you are programming new cluster subprograms, there are additional cluster programmer utilities available; see [CL] **cluster programming utilities** for details.

The cluster dir command provides a directory-style listing of all the recently defined clusters. cluster list provides a detailed listing of the specified clusters, or of all current clusters if no cluster names are specified. The default action is to list all the information attached to the cluster(s). You may limit the type of information listed by specifying particular options.

The cluster drop command removes the named clusters. The keyword _all specifies that all current cluster analyses are to be dropped.

Stata cluster analyses are referenced by name. Many of the cluster commands default to using the most recently defined cluster analysis if no cluster name is provided. The cluster use command places the named cluster analysis as if it were the latest executed cluster analysis, so that, by default, this cluster analysis will be used if the cluster name is omitted from many of the cluster commands. Also realize that you may use the * and ? name-matching characters to shorten the typing of cluster names; see [U] **14.2 Abbreviation rules**.

cluster rename allows you to rename a cluster analysis. This only changes the cluster name. It does not change any of the variable names attached to the cluster analysis. The cluster renamevar command, on the other hand, allows you to rename the variables attached to a cluster analysis and to update the cluster object with the new variable name(s). You do not want to use the rename command (see [R] **rename**) to rename variables attached to a cluster analysis, since this would invalidate the cluster object. Use the cluster renamevar command instead.

Options

all, the default, specifies that all items and information attached to the cluster(s) are to be listed. You may instead pick among the notes, type, method, dissimilarity, similarity, vars, chars, and other options to limit what is presented.

notes specifies that cluster notes are to be listed.

type specifies that the type of cluster analysis is to be listed.

method specifies that the cluster analysis method is to be listed.

dissimilarity specifies that the dissimilarity measure is to be listed.

similarity specifies that the similarity measure is to be listed.

vars specifies that the variables attached to the cluster(s) are to be listed.

chars specifies that any Stata characteristics attached to the cluster(s) are to be listed.

other specifies that information attached to the cluster(s) under the heading "other" are to be listed.

name(clname), for use with cluster renamevar, indicates the cluster analysis within which the variable renaming is to take place. If name() is not specified, the most recently performed cluster analysis (or the one specified by cluster use) will be used.

prefix, for use with cluster renamevar, indicates that all variables attached to the cluster analysis that have oldstub as the beginning of their name are to be renamed, with newstub replacing oldstub.

Remarks

▷ Example

We demonstrate these cluster utility commands by beginning with four already-defined cluster analyses. The dir and list subcommands provide listings of the cluster analyses.

```
. cluster dir
bcx3kmed
ayz5kmeans
abc_clink
xyz_slink
. cluster list xyz_slink
xyz_slink (type: hierarchical,  method: single,  dissimilarity: L2)
        vars: xyz_slink_id (id variable)
              xyz_slink_ord (order variable)
              xyz_slink_hgt (height variable)
       other: range: 0 .
              cmd: cluster singlelinkage x y z, name(xyz_slink)
              varlist: x y z
. cluster list
bcx3kmed (type: partition,  method: kmedians,  dissimilarity: L2)
        vars: bcx3kmed (group variable)
       other: k: 3
              start: krandom
              range: 0 .
              cmd: cluster kmedians b c x, k(3) name(bcx3kmed)
              varlist: b c x
```

```
ayz5kmeans  (type: partition,  method: kmeans,  dissimilarity: L2)
        vars: ayz5kmeans (group variable)
       other: k: 5
              start: krandom
              range: 0 .
              cmd: cluster kmeans a y z, k(5) name(ayz5kmeans)
              varlist: a y z
  abc_clink  (type: hierarchical,  method: complete,  dissimilarity: L2)
        vars: abc_clink_id (id variable)
              abc_clink_ord (order variable)
              abc_clink_hgt (height variable)
       other: range: 0 .
              cmd: cluster completelinkage a b c, name(abc_clink)
              varlist: a b c
  xyz_slink  (type: hierarchical,  method: single,  dissimilarity: L2)
        vars: xyz_slink_id (id variable)
              xyz_slink_ord (order variable)
              xyz_slink_hgt (height variable)
       other: range: 0 .
              cmd: cluster singlelinkage x y z, name(xyz_slink)
              varlist: x y z
. cluster list a*, vars
ayz5kmeans
        vars: ayz5kmeans (group variable)

abc_clink
        vars: abc_clink_id (id variable)
              abc_clink_ord (order variable)
              abc_clink_hgt (height variable)
```

cluster dir listed the names of the four currently defined cluster analyses. cluster list followed by the name of one of the cluster analyses listed the information attached to that cluster analysis. The cluster list command, without an argument, listed the information for all currently defined cluster analyses. We demonstrated the vars option of cluster list to show that you can restrict the information that is listed. Notice also the use of a* as the cluster name. The *, in this case, indicates that any ending is allowed. For these four cluster analyses, it matches the names ayz5kmeans and abc_clink.

We now demonstrate the use of the renamevar subcommand.

```
. cluster renamevar ayz5kmeans g5km
variable ayz5kmeans not found in bcx3kmed
r(198);
. cluster renamevar ayz5kmeans g5km, name(ayz5kmeans)

. cluster list ayz5kmeans
ayz5kmeans  (type: partition,  method: kmeans,  dissimilarity: L2)
        vars: g5km (group variable)
       other: k: 5
              start: krandom
              range: 0 .
              cmd: cluster kmeans a y z, k(5) name(ayz5kmeans)
              varlist: a y z
```

The first use of cluster renamevar failed because we did not specify which cluster object to use (with the name() option), and the most recent cluster object, bcx3kmed, was not the appropriate one. After specifying the name() option with the appropriate cluster name, the renamevar subcommand changed the name as shown in the cluster list command that followed.

The cluster use command places a particular cluster object to be the one used by default. We show this in conjunction with the prefix option of the renamevar subcommand.

```
. cluster use ayz5kmeans

. cluster renamevar g grp, prefix

. cluster renamevar xyz_slink_ wrk, prefix name(xyz*)

. cluster list ayz* xyz*
ayz5kmeans  (type: partition,  method: kmeans,  dissimilarity: L2)
      vars: grp5km (group variable)
     other: k: 5
            start: krandom
            range: 0 .
            cmd: cluster kmeans a y z, k(5) name(ayz5kmeans)
            varlist: a y z
  xyz_slink  (type: hierarchical,  method: single,  dissimilarity: L2)
       vars: wrkid (id variable)
             wrkord (order variable)
             wrkhgt (height variable)
      other: range: 0 .
             cmd: cluster singlelinkage x y z, name(xyz_slink)
             varlist: x y z
```

The cluster use command placed ayz5kmeans as the current cluster object. The cluster renamevar command that followed capitalized on this by leaving off the name() option. The prefix
option allowed the changing of the variable names, as demonstrated in the cluster list of the two
changed cluster objects.

   cluster rename changes the name of cluster objects. cluster drop allows you to drop some
or all of the cluster objects.

```
. cluster rename xyz_slink bob

. cluster rename ayz* sam

. cluster list, type method vars
sam  (type: partition,  method: kmeans)
      vars: grp5km (group variable)
bob  (type: hierarchical,  method: single)
      vars: wrkid (id variable)
            wrkord (order variable)
            wrkhgt (height variable)
bcx3kmed  (type: partition,  method: kmedians)
      vars: bcx3kmed (group variable)
abc_clink  (type: hierarchical,  method: complete)
      vars: abc_clink_id (id variable)
            abc_clink_ord (order variable)
            abc_clink_hgt (height variable)

. cluster drop bcx3kmed abc_clink

. cluster dir
sam
bob

. cluster drop _all

. cluster dir
```

We used options with cluster list to limit what was presented. The _all keyword with cluster
drop removed all currently defined cluster objects.

◁

Also See

Related: [CL] **cluster notes**, [CL] **cluster programming utilities**,
[R] **notes**,
[P] **char**

Background: [CL] **cluster**

Title

cluster wardslinkage — Ward's linkage cluster analysis

Syntax

cluster wardslinkage [*varlist*] [if *exp*] [in *range*] [, name(*clname*)

distance_option generate(*stub*)]

Description

The cluster wardslinkage command performs hierarchical agglomerative Ward's linkage cluster analysis. See [CL] **cluster** for a general discussion of cluster analysis and for a description of the other cluster commands. The cluster dendrogram command (see [CL] **cluster dendrogram**) will display the resulting dendrogram, the cluster stop command (see [CL] **cluster stop**) will help in determining the number of groups, and the cluster generate command (see [CL] **cluster generate**) will produce grouping variables.

Options

name(*clname*) specifies the name to attach to the resulting cluster analysis. If name() is not specified, Stata finds an available cluster name, displays it for your reference, and then attaches the name to your cluster analysis.

distance_option is one of the similarity or dissimilarity measures allowed by Stata. Capitalization of the option does not matter. See [CL] **cluster** for a discussion of these measures.

The available measures designed for continuous data are L2 (synonym Euclidean); L2squared, which is the default for cluster wardslinkage; L1 (synonyms absolute, cityblock, and manhattan); Linfinity (synonym maximum); L(#); Lpower(#): Canberra; correlation; and angular (synonym angle).

The available measures designed for binary data are matching, Jaccard, Russell, Hamman, Dice, antiDice, Sneath, Rogers, Ochiai, Yule, Anderberg, Kulczynski, Gower2, and Pearson.

Several authors advise the exclusive use of the L2squared *distance_option* with Ward's linkage. See the sections *(Dis)similarity transformations and the Lance and Williams formula* and *Warning concerning (dis)similarity choice* in [CL] **cluster** for details.

generate(*stub*) provides a prefix for the variable names created by cluster wardslinkage. By default, the variable-name prefix will be the name specified in name(). Three variables are created and attached to the cluster analysis results, with the suffixes _id, _ord, and _hgt. Users generally will not need to access these variables directly.

Remarks

An example using the default L2squared (squared Euclidean) distance and L2 (Euclidean) distance on continuous data and an example using the matching coefficient on binary data illustrates the cluster wardslinkage command. These are the same datasets introduced in [CL] **cluster singlelinkage**, which are used as examples for all the hierarchical clustering methods, so that you can compare the results from using different hierarchical clustering methods.

▷ Example

As explained in the first example of [CL] **cluster singlelinkage**, as the senior data analyst for a small biotechnology firm, you are given a dataset with 4 chemical laboratory measurements on 50 different samples of a particular plant gathered from the rain forest. The head of the expedition that gathered the samples thinks, based on information from the natives, that an extract from the plant might reduce the negative side effects associated with your company's best-selling nutritional supplement.

While the company chemists and botanists continue exploring the possible uses of the plant and plan future experiments, the head of product development asks you to look at the preliminary data and to report anything that might be helpful to the researchers.

While all 50 of the plants are supposed to be of the same type, you decide to perform a cluster analysis to see if there are subgroups or anomalies among them. Single linkage clustering helped you discover an anomaly in the data. You now wish to see if you discover the same thing using Ward's linkage clustering with the default squared Euclidean distance and with Euclidean distance.

You first call `cluster wardslinkage`, letting the distance default to L2squared (squared Euclidean distance), and use the `name()` option to attach the name `ward` to the resulting cluster analysis. The `cluster list` command (see [CL] **cluster utility**) is then applied to list the components of your cluster analysis. The `cluster dendrogram` command then graphs the dendrogram; see [CL] **cluster dendrogram**. As described in the [CL] **cluster singlelinkage** example, the `labels()` option is used, instead of the default action of showing the observation number, to identify which laboratory technician produced the data.

```
. use http://www.stata-press.com/data/r8/labtech
. cluster wardslinkage x1 x2 x3 x4, name(ward)
. cluster list ward
ward  (type: hierarchical,  method: wards,  dissimilarity: L2squared)
       vars: ward_id (id variable)
             ward_ord (order variable)
             ward_hgt (height variable)
      other: range: 0 .
             cmd: cluster wardslinkage x1 x2 x3 x4, name(ward)
             varlist: x1 x2 x3 x4
. cluster dendrogram ward, vertlab ylab labels(labtech)
```

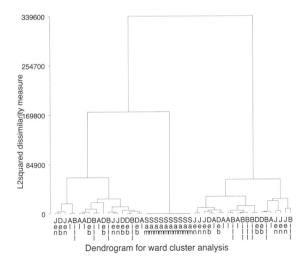

Dendrogram for ward cluster analysis

You now repeat the same analysis using the L2 option to obtain Euclidean distance instead of squared Euclidean distance. This time you name the cluster L2ward.

```
. cluster wardslinkage x1 x2 x3 x4, name(L2ward) L2
. cluster list L2ward
L2ward  (type: hierarchical,  method: wards,  dissimilarity: L2)
      vars: L2ward_id (id variable)
            L2ward_ord (order variable)
            L2ward_hgt (height variable)
     other: range: 0 .
            cmd: cluster wardslinkage x1 x2 x3 x4, name(L2ward) L2
            varlist: x1 x2 x3 x4
. cluster dendrogram L2ward, vertlab ylab labels(labtech)
```

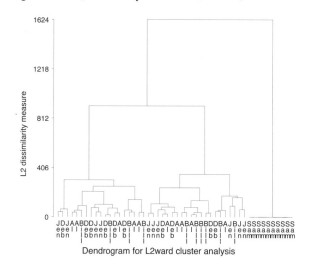

Dendrogram for L2ward cluster analysis

As with single linkage clustering, you see that the samples analyzed by Sam, the lab technician, cluster together closely (dissimilarity measures near zero), and are separated from the rest of the data by a large dissimilarity gap (the long vertical line going up from Sam's cluster to eventually combine with other observations). When you examined the data, you discovered that Sam's data are all between zero and one, while the other four technicians have data that range from zero up to near 150. It appears that Sam has made a mistake.

◁

▷ Example

This example analyzes the same data as introduced in the second example of [CL] **cluster singlelinkage**. The sociology professor of your graduate-level class gives, as homework, a dataset containing 30 observations on 60 binary variables, with the assignment to tell him something about the 30 subjects represented by the observations.

In addition to examining single linkage clustering of these data, you decide to see what Ward's linkage clustering shows. As with the single linkage clustering, you pick the simple matching binary coefficient to measure the similarity between groups. The name() option is used to attach the name wardlink to the cluster analysis. cluster list displays the details; see [CL] **cluster utility**. cluster tree, which is a synonym for cluster dendrogram, then displays the cluster tree (dendrogram); see [CL] **cluster dendrogram**.

```
. use http://www.stata-press.com/data/r8/homework

. cluster ward a1-a60, match name(wardlink)

. cluster list wardlink
wardlink (type: hierarchical,  method: wards,  similarity: matching)
      vars: wardlink_id (id variable)
            wardlink_ord (order variable)
            wardlink_hgt (height variable)
     other: range: 1 0
            cmd: cluster wardslinkage a1-a60, match name(wardlink)
            varlist: a1 a2 a3 a4 a5 a6 a7 a8 a9 a10 a11 a12 a13 a14 a15 a16 a17
                a18 a19 a20 a21 a22 a23 a24 a25 a26 a27 a28 a29 a30 a31 a32
                a33 a34 a35 a36 a37 a38 a39 a40 a41 a42 a43 a44 a45 a46 a47
                a48 a49 a50 a51 a52 a53 a54 a55 a56 a57 a58 a59 a60

. cluster tree wardlink
```

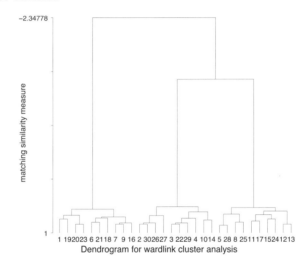

Dendrogram for wardlink cluster analysis

As with single linkage clustering, the dendrogram from Ward's linkage clustering seems to indicate the presence of 3 groups among the 30 observations. However, notice the y-axis range for the resulting dendrogram. How can the matching similarity coefficient range from 1 to less than -2? By definition, the matching coefficient is bounded between 1 and 0. This is an artifact of the way Ward's linkage clustering is defined, and it underscores the warning mentioned in the discussion of the *distance_option*. Also see the sections *(Dis)similarity transformations and the Lance and Williams formula* and *Warning concerning (dis)similarity choice* in [CL] **cluster** for further details.

Later you receive another variable called truegrp that identifies the groups the teacher believes are in the data. You use the cluster generate command (see [CL] **cluster generate**) to create a grouping variable, based on your Ward's linkage clustering, to compare with truegrp. You do a cross-tabulation of truegrp and wardgrp3, your grouping variable, to see if your conclusions match those of the teacher.

(Continued on next page)

```
. cluster gen wardgrp3 = group(3)
. table wardgrp3 truegrp
```

	truegrp		
wardgrp3	1	2	3
1		10	
2	10		
3			10

Other than the numbers arbitrarily assigned to the three groups, your teacher's conclusions and the results from the Ward's linkage clustering are in complete agreement. So, despite the warning against using something other than squared Euclidean distance with Ward's linkage, you were still able to obtain a reasonable cluster analysis solution with the matching similarity coefficient.

◁

❑ Technical Note

cluster wardslinkage requires more memory and more execution time than cluster singlelinkage. With a large number of observations, the execution time may be significant.

❏

Methods and Formulas

[CL] **cluster** discusses hierarchical clustering, and places Ward's linkage clustering in this general framework. The Lance and Williams formula provides the basis for extending the well-known Ward's method of clustering into the general hierarchical linkage framework that allows a choice of (dis)similarity measures.

Conceptually, hierarchical agglomerative Ward's linkage clustering proceeds as follows. The N observations start out as N separate groups each of size one. The two closest observations are merged into one group, producing $N - 1$ total groups. The closest two groups are then merged, so that there are $N - 2$ total groups. This process continues until all the observations are merged into one large group. This produces a hierarchy of groupings from one group to N groups. For Ward's method, the definition of "closest two groups" is based on minimizing the sum of squared errors.

The Ward's linkage clustering algorithm produces two variables that act as a pointer representation of a dendrogram. To this, Stata adds a third variable used to restore the sort order, as needed, so that the two variables of the pointer representation remain valid. The first variable of the pointer representation gives the order of the observations. The second variable has one less element, and gives the height in the dendrogram at which the adjacent observations in the order-variable join.

See [CL] **cluster** for the details, warnings, and formulas of the available *distance_options*, which include (dis)similarity measures for continuous and for binary data.

(Continued on next page)

Also See

Complementary:	[CL] **cluster dendrogram**, [CL] **cluster generate**, [CL] **cluster notes**, [CL] **cluster stop**, [CL] **cluster utility**
Related:	[CL] **cluster averagelinkage**, [CL] **cluster centroidlinkage**, [CL] **cluster completelinkage**, [CL] **cluster medianlinkage**, [CL] **cluster singlelinkage**, [CL] **cluster waveragelinkage**
Background:	[CL] **cluster**

Title

cluster waveragelinkage — Weighted-average linkage cluster analysis

Syntax

cluster <u>w</u>averagelinkage [*varlist*] [if *exp*] [in *range*] [, <u>name</u>(*clname*)

distance_option <u>gene</u>rate(*stub*)]

Description

The cluster waveragelinkage command performs hierarchical agglomerative weighted-average linkage cluster analysis. See [CL] **cluster** for a general discussion of cluster analysis and for a description of the other cluster commands. The cluster dendrogram command (see [CL] **cluster dendrogram**) will display the resulting dendrogram, the cluster stop command (see [CL] **cluster stop**) will help in determining the number of groups, and the cluster generate command (see [CL] **cluster generate**) will produce grouping variables.

Options

name(*clname*) specifies the name to attach to the resulting cluster analysis. If name() is not specified, Stata finds an available cluster name, displays it for your reference, and then attaches the name to your cluster analysis.

distance_option is one of the similarity or dissimilarity measures allowed by Stata. Capitalization of the option does not matter. See [CL] **cluster** for a discussion of these measures.

The available measures designed for continuous data are L2 (synonym <u>Eucl</u>idean), which is the default; L2squared; L1 (synonyms <u>absolute</u>, <u>cityblock</u>, and <u>manhattan</u>); <u>Linf</u>inity (synonym <u>max</u>imum); L(#); <u>Lpower</u>(#); <u>Canb</u>erra; <u>corr</u>elation; and <u>angular</u> (synonym <u>angle</u>).

The available measures designed for binary data are <u>matching</u>, <u>Jac</u>card, <u>Russell</u>, Hamman, Dice, antiDice, Sneath, Rogers, Ochiai, Yule, <u>Ander</u>berg, <u>Kulc</u>zynski, Gower2, and Pearson.

generate(*stub*) provides a prefix for the variable names created by cluster waveragelinkage. By default, the variable-name prefix will be the name specified in name(). Three variables are created and attached to the cluster analysis results, with the suffixes _id, _ord, and _hgt. Users generally will not need to access these variables directly.

Remarks

An example using the default L2 (Euclidean) distance on continuous data and an example using the matching coefficient on binary data illustrate the cluster waveragelinkage command. These are the same datasets introduced in [CL] **cluster singlelinkage**, which are used as examples for all the hierarchical clustering methods, so that you can compare the results from using different hierarchical clustering methods.

▷ Example

As explained in the first example of [CL] **cluster singlelinkage**, as the senior data analyst for a small biotechnology firm, you are given a dataset with 4 chemical laboratory measurements on 50 different samples of a particular plant gathered from the rain forest. The head of the expedition that gathered the samples thinks, based on information from the natives, that an extract from the plant might reduce the negative side effects associated with your company's best-selling nutritional supplement.

While the company chemists and botanists continue exploring the possible uses of the plant and plan future experiments, the head of product development asks you to look at the preliminary data and to report anything that might be helpful to the researchers.

While all 50 of the plants are supposed to be of the same type, you decide to perform a cluster analysis to see if there are subgroups or anomalies among them. Single linkage clustering helped you discover an anomaly in the data. You now wish to see if you discover the same thing using weighted-average linkage clustering with the default Euclidean distance.

You first call `cluster waveragelinkage`, and use the `name()` option to attach the name L2wav to the resulting cluster analysis. The `cluster list` command (see [CL] **cluster utility**) is then applied to list the components of your cluster analysis. The `cluster dendrogram` command then graphs the dendrogram; see [CL] **cluster dendrogram**. As described in the [CL] **cluster singlelinkage** example, the `labels()` option is used, instead of the default action of showing the observation number, to identify which laboratory technician produced the data.

```
. use http://www.stata-press.com/data/r8/labtech
. cluster waveragelinkage x1 x2 x3 x4, name(L2wav)
. cluster list L2wav
L2wav  (type: hierarchical,  method: waverage,  dissimilarity: L2)
        vars: L2wav_id (id variable)
              L2wav_ord (order variable)
              L2wav_hgt (height variable)
       other: range: 0 .
              cmd: cluster waveragelinkage x1 x2 x3 x4, name(L2wav)
              varlist: x1 x2 x3 x4
. cluster dendrogram L2wav, vertlab ylab labels(labtech)
```

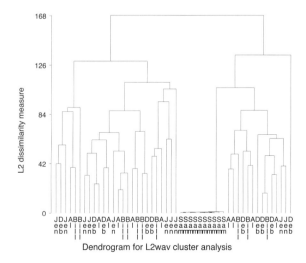

Dendrogram for L2wav cluster analysis

As with single linkage clustering, you see that the samples analyzed by Sam, the lab technician, cluster together closely (dissimilarity measures near zero), and are separated from the rest of the data by a large dissimilarity gap (the long vertical line going up from Sam's cluster to eventually combine with other observations). When you examined the data, you discovered that Sam's data are all between zero and one, while the other four technicians have data that range from zero up to near 150. It appears that Sam has made a mistake.

◁

▷ Example

This example analyzes the same data as introduced in the second example of [CL] **cluster singlelinkage**. The sociology professor of your graduate-level class gives, as homework, a dataset containing 30 observations on 60 binary variables, with the assignment to tell him something about the 30 subjects represented by the observations.

In addition to examining single linkage clustering of these data, you decide to see what weighted-average linkage clustering shows. As with the single linkage clustering, you pick the simple matching binary coefficient to measure the similarity between groups. The name() option is used to attach the name wavlink to the cluster analysis. cluster list displays the details; see [CL] **cluster utility**. cluster tree, which is a synonym for cluster dendrogram, then displays the cluster tree (dendrogram); see [CL] **cluster dendrogram**.

```
. use http://www.stata-press.com/data/r8/homework
. cluster waver a1-a60, match name(wavlink)
. cluster list wavlink
wavlink  (type: hierarchical,  method: waverage,  similarity: matching)
       vars: wavlink_id (id variable)
             wavlink_ord (order variable)
             wavlink_hgt (height variable)
      other: range: 1 0
             cmd: cluster waveragelinkage a1-a60, match name(wavlink)
             varlist: a1 a2 a3 a4 a5 a6 a7 a8 a9 a10 a11 a12 a13 a14 a15 a16 a17
                 a18 a19 a20 a21 a22 a23 a24 a25 a26 a27 a28 a29 a30 a31 a32
                 a33 a34 a35 a36 a37 a38 a39 a40 a41 a42 a43 a44 a45 a46 a47
                 a48 a49 a50 a51 a52 a53 a54 a55 a56 a57 a58 a59 a60
. cluster tree
```

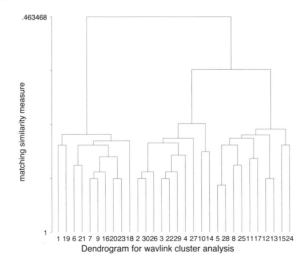

Dendrogram for wavlink cluster analysis

Since, by default, Stata uses the most recently performed cluster analysis, you do not need to type the cluster name when calling `cluster tree`.

As with single linkage clustering, the dendrogram from weighted-average linkage clustering seems to indicate the presence of 3 groups among the 30 observations. Later you receive another variable called `truegrp` that identifies the groups the teacher believes are in the data. You use the `cluster generate` command (see [CL] **cluster generate**) to create a grouping variable, based on your weighted-average linkage clustering, to compare with `truegrp`. You do a cross-tabulation of `truegrp` and `wavgrp3`, your grouping variable, to see if your conclusions match those of the teacher.

```
. cluster gen wavgrp3 = group(3)
. table wavgrp3 truegrp
```

	truegrp		
wavgrp3	1	2	3
1		10	
2	10		
3			10

Other than the numbers arbitrarily assigned to the three groups, your teacher's conclusions and the results from the weighted-average linkage clustering are in complete agreement.

◁

❏ Technical Note

`cluster waveragelinkage` requires more memory and execution time than `cluster singlelinkage`. With a large number of observations, the execution time may be significant.

❏

Methods and Formulas

[CL] **cluster** discusses hierarchical clustering, and places weighted-average linkage clustering in this general framework. Conceptually, hierarchical agglomerative clustering proceeds as follows. The N observations start out as N separate groups each of size one. The two closest observations are merged into one group, producing $N - 1$ total groups. The closest two groups are then merged, so that there are $N - 2$ total groups. This process continues until all the observations are merged into one large group. This produces a hierarchy of groupings from one group to N groups. The difference between the various hierarchical linkage methods depends on how "closest" is defined when comparing groups.

Weighted-average linkage clustering is a variation on average linkage clustering. The difference is in how groups of unequal size are treated. Average linkage gives each observation equal weight. Weighted-average linkage gives each group of observations equal weight, meaning that with unequal group sizes, the observations in the smaller group will have more weight than the observations in the larger group.

The weighted-average linkage clustering algorithm produces two variables that act as a pointer representation of a dendrogram. To this, Stata adds a third variable used to restore the sort order, as needed, so that the two variables of the pointer representation remain valid. The first variable of the pointer representation gives the order of the observations. The second variable has one less element, and gives the height in the dendrogram at which the adjacent observations in the order-variable join.

See [CL] **cluster** for the details, warnings, and formulas of the available *distance_options*, which include (dis)similarity measures for continuous and for binary data.

Also See

Complementary:	[CL] **cluster dendrogram**, [CL] **cluster generate**, [CL] **cluster notes**, [CL] **cluster stop**, [CL] **cluster utility**
Related:	[CL] **cluster averagelinkage**, [CL] **cluster centroidlinkage**, [CL] **cluster completelinkage**, [CL] **cluster medianlinkage**, [CL] **cluster singlelinkage**, [CL] **cluster wardslinkage**
Background:	[CL] **cluster**

Subject and author index

This is the subject and author index for the *Stata Cluster Analysis Reference Manual*. Readers interested in topics other than cluster analysis and graphics should see the combined subject index at the end of Volume 4 of the *Stata Base Reference Manual*, which indexes the *Stata Base Reference Manual*, the *Stata User's Guide*, the *Stata Programming Reference Manual*, the *Stata Cross-Sectional Time-Series Reference Manual*, the *Stata Survey Data Reference Manual*, the *Stata Survival Analysis & Epidemiological Tables Reference Manual*, the *Stata Time-Series Reference Manual*, and this manual. Readers interested in non-cluster-analysis graphics topics should see the index at the end of the *Stata Graphics Reference Manual*.

Semicolons set off the most important entries from the rest. Sometimes no entry will be set off with semicolons; this means all entries are equally important.